NATIONAL ACADEMIES

Foundational Research Gaps and Future Directions for Digital Twins

Committee on Foundational Research Gaps and Future Directions for Digital Twins

Board on Mathematical Sciences and Analytics

Committee on Applied and Theoretical Statistics

Computer Science and Telecommunications Board

Division on Engineering and Physical Sciences

Board on Life Sciences

Board on Atmospheric Sciences and Climate

Division on Earth and Life Studies

National Academy of Engineering

Consensus Study Report

NATIONAL ACADEMIES PRESS 500 Fifth Street, NW Washington, DC 20001

This project was supported by Contract FA9550-22-1-0535 with the Department of Defense (Air Force Office of Scientific Research and Defense Advanced Research Projects Agency), Award Number DE-SC0022878 with the Department of Energy, Award HHSN263201800029I with the National Institutes of Health, and Award AWD-001543 with the National Science Foundation. Any opinions, findings, conclusions, or recommendations expressed in this publication do not necessarily reflect the views of any organization or agency that provided support for the project.

This material is based on work supported by the Department of Energy, Office of Science, Office of Advanced Scientific Computing Research, and Office of Biological and Environmental Research.

This project has been funded in part with federal funds from the National Cancer Institute, National Institute of Biomedical Imaging and Bioengineering, National Library of Medicine, and Office of Data Science Strategy from the National Institutes of Health, Department of Health and Human Services.

This report was prepared as an account of work sponsored by an agency of the United States Government. Neither the United States Government nor any agency thereof, nor any of their employees, makes any warranty, express or implied, or assumes any legal liability or responsibility for the accuracy, completeness, or usefulness of any information, apparatus, product, or process disclosed, or represents that its use would not infringe privately owned rights. Reference herein to any specific commercial product, process, or service by trade name, trademark, manufacturer, or otherwise does not necessarily constitute or imply its endorsement, recommendation, or favoring by the United States Government or any agency thereof. The views and opinions of authors expressed herein do not necessarily state or reflect those of the United States Government or any agency thereof.

International Standard Book Number-13: 978-0-309-70042-9
International Standard Book Number-10: 0-309-70042-6
Digital Object Identifier: https://doi.org/10.17226/26894
Library of Congress Control Number: 2024930543

This publication is available from the National Academies Press, 500 Fifth Street, NW, Keck 360, Washington, DC 20001; (800) 624-6242 or (202) 334-3313; http://www.nap.edu.

Copyright 2024 by the National Academy of Sciences. National Academies of Sciences, Engineering, and Medicine and National Academies Press and the graphical logos for each are all trademarks of the National Academy of Sciences. All rights reserved.

Printed in the United States of America.

Suggested citation: National Academies of Sciences, Engineering, and Medicine. 2024. *Foundational Research Gaps and Future Directions for Digital Twins*. Washington, DC: The National Academies Press. https://doi.org/10.17226/26894.

The **National Academy of Sciences** was established in 1863 by an Act of Congress, signed by President Lincoln, as a private, nongovernmental institution to advise the nation on issues related to science and technology. Members are elected by their peers for outstanding contributions to research. Dr. Marcia McNutt is president.

The **National Academy of Engineering** was established in 1964 under the charter of the National Academy of Sciences to bring the practices of engineering to advising the nation. Members are elected by their peers for extraordinary contributions to engineering. Dr. John L. Anderson is president.

The **National Academy of Medicine** (formerly the Institute of Medicine) was established in 1970 under the charter of the National Academy of Sciences to advise the nation on medical and health issues. Members are elected by their peers for distinguished contributions to medicine and health. Dr. Victor J. Dzau is president.

The three Academies work together as the **National Academies of Sciences, Engineering, and Medicine** to provide independent, objective analysis and advice to the nation and conduct other activities to solve complex problems and inform public policy decisions. The National Academies also encourage education and research, recognize outstanding contributions to knowledge, and increase public understanding in matters of science, engineering, and medicine.

Learn more about the National Academies of Sciences, Engineering, and Medicine at **www.nationalacademies.org**.

Consensus Study Reports published by the National Academies of Sciences, Engineering, and Medicine document the evidence-based consensus on the study's statement of task by an authoring committee of experts. Reports typically include findings, conclusions, and recommendations based on information gathered by the committee and the committee's deliberations. Each report has been subjected to a rigorous and independent peer-review process and it represents the position of the National Academies on the statement of task.

Proceedings published by the National Academies of Sciences, Engineering, and Medicine chronicle the presentations and discussions at a workshop, symposium, or other event convened by the National Academies. The statements and opinions contained in proceedings are those of the participants and are not endorsed by other participants, the planning committee, or the National Academies.

Rapid Expert Consultations published by the National Academies of Sciences, Engineering, and Medicine are authored by subject-matter experts on narrowly focused topics that can be supported by a body of evidence. The discussions contained in rapid expert consultations are considered those of the authors and do not contain policy recommendations. Rapid expert consultations are reviewed by the institution before release.

For information about other products and activities of the National Academies, please visit www.nationalacademies.org/about/whatwedo.

COMMITTEE ON FOUNDATIONAL RESEARCH GAPS AND FUTURE DIRECTIONS FOR DIGITAL TWINS

KAREN E. WILLCOX (NAE), The University of Texas at Austin, *Chair*
DEREK BINGHAM, Simon Fraser University
CAROLINE CHUNG, MD Anderson Cancer Center
JULIANNE CHUNG, Emory University
CAROLINA CRUZ-NEIRA (NAE), University of Central Florida
CONRAD J. GRANT, Johns Hopkins University Applied Physics Laboratory
JAMES L. KINTER, George Mason University
RUBY LEUNG (NAE), Pacific Northwest National Laboratory
PARVIZ MOIN (NAS/NAE), Stanford University
LUCILA OHNO-MACHADO (NAM), Yale University
COLIN J. PARRIS (NAE), GE Vernova Digital
IRENE QUALTERS, Los Alamos National Laboratory
INES THIELE, University of Galway
CONRAD TUCKER, Carnegie Mellon University
REBECCA WILLETT, The University of Chicago
XINYUE YE, Texas A&M University

Staff

BRITTANY SEGUNDO, Program Officer, Board on Mathematical Sciences and Analytics (BMSA), *Study Director*
KAVITA BERGER, Director, Board on Life Sciences
ELIZABETH T. CADY, Senior Program Officer, National Academy of Engineering
JON EISENBERG, Director, Computer Science and Telecommunications Board (CSTB)
SAMANTHA KORETSKY, Research Assistant, BMSA
PADMA LIM, College Student Intern, BMSA
HEATHER LOZOWSKI, Senior Finance Business Partner
THO NGUYEN, Senior Program Officer, CSTB
JOE PALMER, Senior Project Assistant, BMSA
PATRICIA RAZAFINDRAMBININA, Associate Program Officer, Board on Atmospheric Sciences and Climate (until April 2023)
BLAKE REICHMUTH, Associate Program Officer, BMSA
MICHELLE K. SCHWALBE, Director, BMSA
ERIK SVEDBERG, Scholar, National Materials and Manufacturing Board
NNEKA UDEAGBALA, Associate Program Officer, CSTB

BOARD ON MATHEMATICAL SCIENCES AND ANALYTICS

MARK L. GREEN, University of California, Los Angeles, *Co-Chair*
KAREN E. WILLCOX (NAE), The University of Texas at Austin, *Co-Chair*
HÉLÈNE BARCELO, Mathematical Sciences Research Institute
BONNIE BERGER (NAS), Massachusetts Institute of Technology
RUSSEL E. CAFLISCH (NAS), New York University
DAVID S.C. CHU, Institute for Defense Analyses
DUANE COOPER, Morehouse College
JAMES H. CURRY, University of Colorado Boulder
RONALD D. FRICKER, JR., Virginia Polytechnic Institute and State University
JULIE IVY, North Carolina State University
LYDIA E. KAVRAKI (NAM), Rice University
TAMARA G. KOLDA (NAE), Sandia National Laboratories
PETROS KOUMOUTSAKOS (NAE), Harvard University
RACHEL KUSKE, Georgia Institute of Technology
YANN A. LeCUN (NAS/NAE), Facebook
JILL C. PIPHER, Brown University
YORAM SINGER, WorldQuant
TATIANA TORO, University of Washington
JUDY WALKER, University of Nebraska–Lincoln
LANCE A. WALLER, Emory University

Staff

MICHELLE K. SCHWALBE, Director
SAMANTHA KORETSKY, Research Assistant
HEATHER LOZOWSKI, Senior Finance Business Partner
JOE PALMER, Senior Project Assistant
BLAKE REICHMUTH, Associate Program Officer
BRITTANY SEGUNDO, Program Officer

COMMITTEE ON APPLIED AND THEORETICAL STATISTICS

ELIZABETH A. STUART, Johns Hopkins Bloomberg School of Public Health, *Co-Chair*
LANCE A. WALLER, Emory University, *Co-Chair*
FREDERICK D. BOWMAN (NAM), University of Michigan
WEI CHEN (NAE), Northwestern University
OMAR GHATTAS, The University of Texas at Austin
OFER HAREL, University of Connecticut
TIM HESTERBERG, Instacart
REBECCA A. HUBBARD, University of Pennsylvania
KRISTIN LAUTER, Meta
MIGUEL MARINO (NAM), Oregon Health & Science University
XIAO-LI MENG, Harvard University
BHRAMAR MUKHERJEE (NAM), University of Michigan
RAQUEL PRADO, University of California, Santa Cruz
KIMBERLY FLAGG SELLERS, North Carolina State University
PIYUSHIMITA THAKURIAH, Rutgers University
BRANI VIDAKOVIC, Texas A&M University
JONATHAN WAKEFIELD, University of Washington
YAZHEN WANG, University of Wisconsin–Madison
ALYSON G. WILSON, North Carolina State University

Staff

BRITTANY SEGUNDO, Director
SAMANTHA KORETSKY, Research Assistant
HEATHER LOZOWSKI, Senior Finance Business Partner
JOE PALMER, Senior Project Assistant
BLAKE REICHMUTH, Associate Program Officer
MICHELLE SCHWALBE, Director, Board on Mathematical Sciences and Analytics

COMPUTER SCIENCE AND TELECOMMUNICATIONS BOARD

LAURA M. HAAS (NAE), University of Massachusetts Amherst, *Chair*
DAVID E. CULLER (NAE), University of California, Berkeley
ERIC HORVITZ (NAE), Microsoft Research Lab–Redmond
CHARLES ISBELL, Georgia Institute of Technology
ELIZABETH MYNATT, Georgia Institute of Technology
CRAIG PARTRIDGE, BBN Corporation
DANIELA RUS (NAE), Massachusetts Institute of Technology
FRED B. SCHNEIDER (NAE), Cornell University
MARGO I. SELTZER (NAE), University of British Columbia
NAMBIRAJAN SESHADRI (NAE), University of California, San Diego
MOSHE Y. VARDI (NAS/NAE), Rice University

Staff

JON EISENBERG, Senior Board Director
SHENAE BRADLEY, Administrative Assistant
RENEE HAWKINS, Financial and Administrative Manager
THO NGUYEN, Senior Program Officer
GABRIELLE RISICA, Program Officer
BRENDAN ROACH, Program Officer
NNEKA UDEAGBALA, Associate Program Officer

BOARD ON LIFE SCIENCES

ANN M. ARVIN (NAM), Stanford University, *Chair*
BARBARA A. SCHAAL (NAS),[1] Washington University in St. Louis, *Chair*
A. ALONSO AGUIRRE,[1] George Mason University
DENISE N. BAKEN, Shield Analysis Technology, LLC
VALERIE H. BONHAM, Kennedy Krieger Institute
PATRICK M. BOYLE, Ginkgo Bioworks, Inc.
DOMINIQUE BROSSARD, University of Wisconsin–Madison
SCOTT V. EDWARDS (NAS),[1] Harvard University
GERALD L. EPSTEIN, Johns Hopkins Center for Health Security
ROBERT J. FULL,[1] University of California, Berkeley
BERONDA MONTGOMERY, Michigan State University
LOUIS J. MUGLIA (NAM), Burroughs Wellcome Fund
ROBERT NEWMAN, Aspen Institute
LUCILA OHNO-MACHADO (NAM), Yale University
SUDIP S. PARIKH, American Association for the Advancement of Science
NATHAN D. PRICE, Institute for Systems Biology
SUSAN R. SINGER, St. Olaf College
DAVID R. WALT, Harvard Medical School
PHYLLIS M. WISE (NAM), University of Illinois at Urbana-Champaign

Staff

KAVITA M. BERGER, Director
ANDREW BREMER, Program Officer
NANCY CONNELL, Senior Scientist
JESSICA DE MOUY, Research Associate
CYNTHIA GETNER, Senior Financial Business Partner
LYLY LUHACHACK, Program Officer
DASIA McKOY, Program Assistant
STEVEN MOSS, Senior Program Officer
CHRISTL SAUNDERS, Program Coordinator
AUDREY THEVENON, Senior Program Officer
TRISHA TUCHOLSKI, Associate Program Officer
SABINA VADNAIS, Research Associate
NAM VU, Program Assistant

---

[1] Committee membership term ended on June 30, 2023.

BOARD ON ATMOSPHERIC SCIENCES AND CLIMATE

MARY GLACKIN, The Weather Company, an IBM Business, *Chair*
CYNTHIA S. ATHERTON, Heising-Simons Foundation
ELIZABETH A. BARNES, Colorado State University
BRAD R. COLMAN, The Climate Corporation
BARTHOLOMEW E. CROES, California Air Resources Board
NEIL DONAHUE, Carnegie Mellon University
ROBERT B. DUNBAR, Stanford University
LESLEY-ANN DUPIGNY-GIROUX, University of Vermont
EFI FOUFOULA-GEORGIOU (NAE), University of California, Irvine
PETER C. FRUMHOFF, Harvard University
ROBERT KOPP, Rutgers University
RUBY LEUNG (NAE), Pacific Northwest National Laboratory
ZHANQING LI, University of Maryland
JONATHAN MARTIN, University of Wisconsin–Madison
AMY McGOVERN, Oklahoma State University
LINDA O. MEARNS, National Center for Atmospheric Research
JONATHAN A. PATZ (NAM), University of Wisconsin–Madison
J. MARSHALL SHEPHERD (NAS/NAE), University of Georgia
DAVID W. TITLEY, The Pennsylvania State University
ARADHNA TRIPATI, University of California, Los Angeles
ELKE U. WEBER (NAS), Princeton University

Staff

STEVEN STICHTER, Interim Board Director
KYLE ALDRIDGE, Senior Program Assistant
APURVA DAVE, Senior Program Officer
RITA GASKINS, Administrative Coordinator
ROB GREENWAY, Program Associate
KATRINA HUI, Associate Program Officer
BRIDGET McGOVERN, Associate Program Officer
APRIL MELVIN, Senior Program Officer
AMY MITSUMORI, Research Associate
LINDSAY MOLLER, Senior Program Assistant
MORGAN MONZ, Associate Program Officer
AMANDA PURCELL, Senior Program Officer
PATRICIA RAZAFINDRAMBININA, Associate Program Officer (until April 2023)
ALEX REICH, Program Officer
RACHEL SANCHEZ, Program Assistant
RACHEL SILVERN, Program Officer
HUGH WALPOLE, Associate Program Officer

Reviewers

This Consensus Study Report was reviewed in draft form by individuals chosen for their diverse perspectives and technical expertise. The purpose of this independent review is to provide candid and critical comments that will assist the National Academies of Sciences, Engineering, and Medicine in making each published report as sound as possible and to ensure that it meets the institutional standards for quality, objectivity, evidence, and responsiveness to the study charge. The review comments and draft manuscript remain confidential to protect the integrity of the deliberative process.

We thank the following individuals for their review of this report:

BRENT APPLEBY, Draper
PETER BAUER, European Centre for Medium-Range Weather Forecasts (retired)
RUSSEL E. CAFLISCH (NAS), New York University
WEI CHEN (NAE), Northwestern University
JACK DEMPSEY, Asset Management Partnership, LLC
ALBERTO FERRARI, RTX
MISHA KILMER, Tufts University
AMINA QUTUB, The University of Texas at San Antonio
DAWN TILBURY, University of Michigan

Although the reviewers listed above provided many constructive comments and suggestions, they were not asked to endorse the conclusions or recommendations of this report nor did they see the final draft before its release. The review of this report was overseen by **ALICIA L. CARRIQUIRY (NAM)**, Iowa State University, and **ROBERT F. SPROULL (NAE)**, University of Massachusetts Amherst. They were responsible for making certain that an independent examination of this report was carried out in accordance with the standards of the National Academies and that all review comments were carefully considered. Responsibility for the final content rests entirely with the authoring committee and the National Academies.

Acknowledgments

We are grateful to the many scholars and leaders who contributed their time and expertise to the committee's information-gathering efforts. Special thanks go to all the speakers who briefed the committee: Chaitanya Baru, National Science Foundation; Tim Booher, Boeing; Steve Dennis, International Computer Science Institute; Alberto Ferrari, RTX; Omar Ghattas, The University of Texas at Austin; Mark Girolami, The Alan Turing Institute; Philip Huff, University of Arkansas at Little Rock; George Karniadakis, Brown University; Michael Mahoney, University of California, Berkeley; Johannes Royset, Naval Postgraduate School; Lea Shanley, International Computer Science Institute; and Jack Wells, NVIDIA.

We are equally appreciative of all the panelists and speakers in our workshop series.

- **Workshop on Opportunities and Challenges for Digital Twins in Biomedical Sciences:** Bissan Al-Lazikani, MD Anderson Cancer Center; Rommie Amaro, University of California, San Diego; Gary An, University of Vermont; Aldo Badano, Food and Drug Administration; Mikael Benson, Karolinska Institute; Todd Coleman, Stanford University; Heiko Enderling, H. Lee Moffitt Cancer Center; Liesbet Geris, University of Liège; James A. Glazier, Indiana University; Jayashree Kalpathy-Cramer, University of Colorado Denver; Petros Koumoutsakos, Harvard University; Reinhard Laubenbacher, University of Florida; Lara Mangravite, HI-Bio; David Miller, Unlearn.AI; Juan Perilla, University of Delaware; Jodyn Platt, University of Michigan; Nathan Price, Thorne HealthTech; Jeffrey R. Sachs, Merck & Co., Inc.; Karissa Sanbonmatsu, Los Alamos

National Laboratory; and Tom Yankeelov, The University of Texas at Austin.
- **Workshop on Digital Twins in Atmospheric, Climate, and Sustainability Science:** Anima Anandkumar, California Institute of Technology; Mark Asch, Sandia National Laboratories; Venkatramani Balaji, Schmidt Futures; Elizabeth A. Barnes, Colorado State University; Cecilia Bitz, University of Washington; Emanuele Di Lorenzo, Brown University; Omar Ghattas, The University of Texas at Austin; Mike Goodchild, University of California, Santa Barbara; John Harlim, The Pennsylvania State University; Christiane Jablonowski, University of Michigan; Jean-Francois Lamarque, McKinsey; Amy McGovern, University of Oklahoma; Anna Michalak, Carnegie Institution for Science; Umberto Modigliani, European Centre for Medium-Range Weather Forecasts; Mike Pritchard, NVIDIA/University of California, Irvine; Abhinav Saxena, GE Research; Gavin A. Schmidt, National Aeronautics and Space Administration Goddard Institute for Space Studies; Tapio Schneider, California Institute of Technology; Mark Taylor, Sandia National Laboratories; and Yuyu Zhou, Iowa State University.
- **Workshop on Opportunities and Challenges for Digital Twins in Engineering:** Elizabeth Baron, Unity Technologies; Grace Bochenek, University of Central Florida; José R. Celaya, Schlumberger; Dinakar Deshmukh, General Electric; Karthik Duraisamy, University of Michigan; Charles Farrar, Los Alamos National Laboratory; Devin Francom, Los Alamos National Laboratory; S. Michael Gahn, Rolls-Royce; Michael Grieves, Digital Twin Institute; Devin Harris, University of Virginia; and Pamela Kobryn, Department of Defense.

These workshops and this report were the products of the committee's thoughtful deliberation and dedication to the topic. We extend our gratitude to the broader community for their engagement with this project. Finally, we appreciate the collaborative efforts of every member of the staff team.

Contents

SUMMARY		1
1	INTRODUCTION	17
2	THE DIGITAL TWIN LANDSCAPE	21
3	VIRTUAL REPRESENTATION: FOUNDATIONAL RESEARCH NEEDS AND OPPORTUNITIES	49
4	THE PHYSICAL COUNTERPART: FOUNDATIONAL RESEARCH NEEDS AND OPPORTUNITIES	69
5	FEEDBACK FLOW FROM PHYSICAL TO VIRTUAL: FOUNDATIONAL RESEARCH NEEDS AND OPPORTUNITIES	78
6	FEEDBACK FLOW FROM VIRTUAL TO PHYSICAL: FOUNDATIONAL RESEARCH NEEDS AND OPPORTUNITIES	85
7	TOWARD SCALABLE AND SUSTAINABLE DIGITAL TWINS	99
8	SUMMARY OF FINDINGS, CONCLUSIONS, AND RECOMMENDATIONS	114

APPENDIXES

A	Statement of Task	127
B	Workshop Agendas	129
C	Opportunities and Challenges for Digital Twins in Atmospheric and Climate Sciences: Proceedings of a Workshop—in Brief	135
D	Opportunities and Challenges for Digital Twins in Biomedical Research: Proceedings of a Workshop—in Brief	149
E	Opportunities and Challenges for Digital Twins in Engineering: Proceedings of a Workshop—in Brief	163
F	Committee Member Biographical Information	177
G	Acronyms and Abbreviations	186

Summary

DEFINITION OF A DIGITAL TWIN

The Committee on Foundational Research Gaps and Future Directions for Digital Twins uses the following definition of a digital twin, modified from a definition published by the American Institute of Aeronautics and Astronautics (AIAA Digital Engineering Integration Committee 2020):

> A digital twin is a set of virtual information constructs that mimics the structure, context, and behavior of a natural, engineered, or social system (or system-of-systems), is dynamically updated with data from its physical twin, has a predictive capability, and informs decisions that realize value. The bidirectional interaction between the virtual and the physical is central to the digital twin.

The study committee's refined definition refers to "a natural, engineered, or social system (or system-of-systems)" to describe digital twins of physical systems in the broadest sense possible, including the engineered world, natural phenomena, biological entities, and social systems. This definition introduces the phrase "predictive capability" to emphasize that a digital twin must be able to issue predictions beyond the available data to drive decisions that realize value. Finally, this definition highlights the bidirectional interaction, which comprises feedback flows of information from the physical system to the virtual representa-

---

NOTE: This summary highlights key messages from the report but is not exhaustive. In order to support the flow and readability of this abridged summary, the findings, conclusions, and recommendations may be ordered differently than in the main body of the report, but they retain the same numbering scheme for searchability.

tion and from the virtual back to the physical system to enable decision-making, either automatic or with humans-in-the-loop.

THE PROMISE OF DIGITAL TWINS

Digital twins hold immense promise in accelerating scientific discovery and revolutionizing industries. Digital twins can be a critical tool for decision-making based on a synergistic combination of models and data. The bidirectional interplay between a physical system and its virtual representation endows the digital twin with a dynamic nature that goes beyond what has been traditionally possible with modeling and simulation, creating a virtual representation that evolves with the system over time. By enabling predictive insights and effective optimizations, monitoring performance to detect anomalies and exceptional conditions, and simulating dynamic system behavior, digital twins have the capacity to revolutionize scientific research, enhance operational efficiency, optimize production strategies, reduce time-to-market, and unlock new avenues for scientific and industrial growth and innovation. The use cases for digital twins are diverse and proliferating, with applications across multiple areas of science, technology, and society, and their potential is wide-reaching. Yet key research needs remain to advance digital twins in several domains.

This report is the result of a study that addressed the following key topics:

- Definitions of and use cases for digital twins;
- Foundational mathematical, statistical, and computational gaps for digital twins;
- Best practices for digital twin development and use; and
- Opportunities to advance the use and practice of digital twins.

While there is significant enthusiasm around industry developments and applications of digital twins, the focus of this report is on identifying research gaps and opportunities. The report's recommendations are particularly targeted toward what agencies and researchers can do to advance mathematical, statistical, and computational foundations of digital twins.

ELEMENTS OF THE DIGITAL TWIN ECOSYSTEM

The notion of a digital twin goes beyond simulation to include tighter integration between models, data, and decisions. The dynamic, bidirectional interaction tailors the digital twin to a particular physical counterpart and supports the evolution of the virtual representation as the physical counterpart evolves. This bidirectional interaction is sometimes characterized as a feedback loop, where data from the physical counterpart are used to update the virtual models, and, in turn, the virtual models are used to drive changes in the physical system.

This feedback loop may occur in real time, such as for dynamic control of an autonomous vehicle or a wind farm, or it may occur on slower time scales, such as post-flight updating of a digital twin for aircraft engine predictive maintenance or post-imaging updating of a digital twin and subsequent treatment planning for a cancer patient.

The digital twin provides decision support when a human plays a decision-making role, or decision-making may be shared jointly between the digital twin and a human as a human–agent team. Human–digital twin interactions may also involve the human playing a crucial role in designing, managing, and operating elements of the digital twin, including selecting sensors and data sources, managing the models underlying the virtual representation, and implementing algorithms and analytics tools.

Finding 2-1: A digital twin is more than just simulation and modeling.

Conclusion 2-1: The key elements that comprise a digital twin include (1) modeling and simulation to create a virtual representation of a physical counterpart, and (2) a bidirectional interaction between the virtual and the physical. This bidirectional interaction forms a feedback loop that comprises dynamic data-driven model updating (e.g., sensor fusion, inversion, data assimilation) and optimal decision-making (e.g., control, sensor steering).

These elements are depicted in Figure S-1.

An important theme that runs throughout this report is the notion that the digital twin virtual representation be "fit for purpose," meaning that the virtual representation—model types, fidelity, resolution, parameterization, and quantities of interest—be chosen, and in many cases dynamically adapted, to fit the particular decision task and computational constraints at hand.

Conclusion 3-1: A digital twin should be defined at a level of fidelity and resolution that makes it fit for purpose. Important considerations are the required level of fidelity for prediction of the quantities of interest, the available computational resources, and the acceptable cost. This may lead to the digital twin including high-fidelity, simplified, or surrogate models, as well as a mixture thereof. Furthermore, a digital twin may include the ability to represent and query the virtual models at variable levels of resolution and fidelity depending on the particular task at hand and the available resources (e.g., time, computing, bandwidth, data).

An additional consideration is the complementary role of models and data—a digital twin is distinguished from traditional modeling and simulation in the way that models and data work together to drive decision-making. In cases in

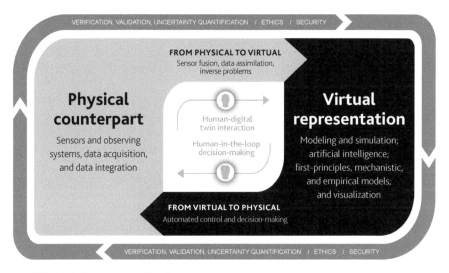

FIGURE S-1 Elements of the digital twin ecosystem.
NOTES: Information flows bidirectionally between the virtual representation and physical counterpart. These information flows may be through automated processes, human-driven processes, or a combination of the two.

which an abundance of data exists and the decisions to be made fall largely within the realm of conditions represented by the data, a data-centric view of a digital twin is appropriate—the data form the core of the digital twin, the numerical model is likely heavily empirical, and analytics and decision-making wrap around this numerical model. In other cases that are data-poor and call on the digital twin to issue predictions in extrapolatory regimes that go well beyond the available data, a model-centric view of a digital twin is appropriate—a mathematical model and its associated numerical model form the core of the digital twin, and data are assimilated through the lens of these models. An important need is to advance hybrid modeling approaches that leverage the synergistic strengths of data-driven and model-driven digital twin formulations.

ADVANCING DIGITAL TWIN STATE OF THE ART REQUIRES AN INTEGRATED RESEARCH AGENDA

Despite the existence of examples of digital twins providing practical impact and value, the sentiment expressed across multiple committee information-gathering sessions is that the publicity around digital twins and digital twin solutions currently outweighs the evidence base of success.

Conclusion 2-5: Digital twins have been the subject of widespread interest and enthusiasm; it is challenging to separate what is true from what is merely aspirational, due to a lack of agreement across domains and sectors as well as misinformation. It is important to separate the aspirational from the actual to strengthen the credibility of the research in digital twins and to recognize that serious research questions remain in order to achieve the aspirational.

Conclusion 2-6: Realizing the potential of digital twins requires an integrated research agenda that advances each one of the key digital twin elements and, importantly, a holistic perspective of their interdependencies and interactions. This integrated research agenda includes foundational needs that span multiple domains as well as domain-specific needs.

Recommendation 1: Federal agencies should launch new crosscutting programs, such as those listed below, to advance mathematical, statistical, and computational foundations for digital twins. As these new digital twin–focused efforts are created and launched, federal agencies should identify opportunities for cross-agency interactions and facilitate cross-community collaborations where fruitful. An interagency working group may be helpful to ensure coordination.

- *National Science Foundation (NSF).* **NSF should launch a new program focused on mathematical, statistical, and computational foundations for digital twins that cuts across multiple application domains of science and engineering.**
    - **The scale and scope of this program should be in line with other multidisciplinary NSF programs (e.g., NSF Artificial Intelligence Institutes) to highlight the technical challenge being solved as well as the emphasis on theoretical foundations being grounded in practical use cases.**
    - **Ambitious new programs launched by NSF for digital twins should ensure that sufficient resources are allocated to the solicitation so that the technical advancements are evaluated using real-world use cases and testbeds.**
    - **NSF should encourage collaborations across industry and academia and develop mechanisms to ensure that small and medium-sized industrial and academic institutions can also compete and be successful leading such initiatives.**
    - **Ideally, this program should be administered and funded by multiple directorates at NSF, ensuring that from inception to sunset, real-world applications in multiple domains guide the theoretical components of the program.**

- *Department of Energy (DOE).* DOE should draw on its unique computational facilities and large instruments coupled with the breadth of its mission as it considers new crosscutting programs in support of digital twin research and development. It is well positioned and experienced in large, interdisciplinary, multi-institutional mathematical, statistical, and computational programs. Moreover, it has demonstrated the ability to advance common foundational capabilities while also maintaining a focus on specific use-driven requirements (e.g., predictive high-fidelity models for high-consequence decision support). This collective ability should be reflected in a digital twin grand challenge research and development vision for DOE that goes beyond the current investments in large-scale simulation to advance and integrate the other digital twin elements, including the physical/virtual bidirectional interaction and high-consequence decision support. This vision, in turn, should guide DOE's approach in establishing new crosscutting programs in mathematical, statistical, and computational foundations for digital twins.
- *National Institutes of Health (NIH).* NIH should invest in filling the gaps in digital twin technology in areas that are particularly critical to biomedical sciences and medical systems. These include bioethics, handling of measurement errors and temporal variations in clinical measurements, capture of adequate metadata to enable effective data harmonization, complexities of clinical decision-making with digital twin interactions, safety of closed-loop systems, privacy, and many others. This could be done via new cross-institute programs and expansion of current programs such as the Interagency Modeling and Analysis Group.
- *Department of Defense (DoD).* DoD's Office of the Under Secretary of Defense for Research and Engineering should advance the application of digital twins as an integral part of the digital engineering performed to support system design, performance analysis, developmental and operational testing, operator and force training, and operational maintenance prediction. DoD should also consider using mechanisms such as the Multidisciplinary University Research Initiative and Defense Acquisition University to support research efforts to develop and mature the tools and techniques for the application of digital twins as part of system digital engineering and model-based system engineering processes.
- *Other federal agencies.* Many federal agencies and organizations beyond those listed above can play important roles in the advancement of digital twin research. For example, the National Oceanic and Atmospheric Administration, the National Institute of Standards and Technology, and the National Aeronautics and Space Administration

should be included in the discussion of digital twin research and development, drawing on their unique missions and extensive capabilities in the areas of data assimilation and real-time decision support.

Verification, Validation, and Uncertainty Quantification: Foundational Research Needs and Opportunities

Verification, validation, and uncertainty quantification (VVUQ) is an area of particular need that necessitates collaborative and interdisciplinary investment to advance the responsible development, implementation, monitoring, and sustainability of digital twins. Evolution of the physical counterpart in real-world use conditions, changes in data collection, noisiness of data, addition and deletion of data sources, changes in the distribution of the data shared with the virtual twin, changes in the prediction and/or decision tasks posed to the digital twin, and evolution of the digital twin virtual models all have consequences for VVUQ.

VVUQ must play a role in all elements of the digital twin ecosystem. In the digital twin virtual representation, verification and validation play key roles in building trustworthiness, while uncertainty quantification gives measures of the quality of prediction. Many of the elements of VVUQ for digital twins are shared with VVUQ for computational models (NRC 2012), although digital twins bring some additional challenges. Common challenges arise from model discrepancies, unresolved scales, surrogate modeling, and the need to issue predictions in extrapolatory regimes. However, digital twin VVUQ must also address the uncertainties associated with the physical counterpart, including changes to sensors or data collection equipment, and the continual evolution of the physical counterpart's state. Data quality improvements may be prioritized based on the relative impacts of parameter uncertainties on the model uncertainties. VVUQ also plays a role in understanding the impact of mechanisms used to pass information between the physical and virtual. These include challenges arising from parameter uncertainty and ill-posed or indeterminate inverse problems, in addition to the uncertainty introduced by the inclusion of the human-in-the-loop.

Conclusion 2-2: Digital twins require VVUQ to be a continual process that must adapt to changes in the physical counterpart, digital twin virtual models, data, and the prediction/decision task at hand. A gap exists between the class of problems that has been considered in traditional modeling and simulation settings and the VVUQ problems that will arise for digital twins.

Conclusion 2-3: Despite the growing use of artificial intelligence, machine learning, and empirical modeling in engineering and scientific applications, there is a lack of standards in reporting VVUQ as well as a lack of consideration of confidence in modeling outputs.

Conclusion 2-4: Methods for ensuring continual VVUQ and monitoring of digital twins are required to establish trust. It is critical that VVUQ be deeply embedded in the design, creation, and deployment of digital twins. In future digital twin research developments, VVUQ should play a core role and tight integration should be emphasized. Particular areas of research need include continual verification, continual validation, VVUQ in extrapolatory conditions, and scalable algorithms for complex multiscale, multiphysics, and multi-code digital twin software efforts. There is a need to establish to what extent VVUQ approaches can be incorporated into automated online operations of digital twins and where new approaches to online VVUQ may be required.

Recommendation 2: Federal agencies should ensure that verification, validation, and uncertainty quantification (VVUQ) is an integral part of new digital twin programs. In crafting programs to advance the digital twin VVUQ research agenda, federal agencies should pay attention to the importance of (1) overarching complex multiscale, multiphysics problems as catalysts to promote interdisciplinary cooperation; (2) the availability and effective use of data and computational resources; (3) collaborations between academia and mission-driven government laboratories and agencies; and (4) opportunities to include digital twin VVUQ in educational programs. Federal agencies should consider the Department of Energy Predictive Science Academic Alliance Program as a possible model to emulate.

Virtual Representation: Foundational Research Needs and Opportunities

A fundamental challenge for digital twins is the vast range of spatial and temporal scales that the virtual representation may need to address. In many applications, a gap remains between the scales that can be simulated and actionable scales. An additional challenge is that as finer scales are resolved and a given model achieves greater fidelity to the physical counterpart it simulates, the computational and data storage/analysis requirements increase. This limits the applicability of the model for some purposes, such as uncertainty quantification, probabilistic prediction, scenario testing, and visualization.

Finding 3-2: Different applications of digital twins drive different requirements for modeling fidelity, data, precision, accuracy, visualization, and time-to-solution, yet many of the potential uses of digital twins are currently intractable to realize with existing computational resources.

Recommendation 3: In crafting research programs to advance the foundations and applications of digital twins, federal agencies should create mechanisms to provide digital twin researchers with computational resources, recognizing the large existing gap between simulated and actionable scales and the differing levels of maturity of high-performance computing across communities.

Mathematical and algorithmic advances in data-driven modeling and multiscale physics-based modeling are necessary elements for closing the gap between simulated and actionable scales. Reductions in computational and data requirements achieved through algorithmic advances are an important complement to increased computing resources. Important areas to advance include hybrid modeling approaches—a synergistic combination of empirical and mechanistic modeling approaches that leverage the best of both data-driven and model-driven formulations—and surrogate modeling approaches. Key gaps, research needs, and opportunities include the following:

- Combining data-driven models with mechanistic models requires effective coupling techniques to facilitate the flow of information (data, variables, etc.) between the models while understanding the inherent constraints and assumptions of each model.
- Integration of component/subsystem digital twins is a pacing item for the digital twin representation of a complex system, especially if different fidelity models are used in the representation of its components/subsystems. There are key gaps in quantifying the uncertainty in digital twins of coupled complex systems, enhancing interoperability between digital twin models, and reconciling assumptions made between models.
- Methods are needed to achieve VVUQ of hybrid and surrogate models, recognizing the uncertain conditions under which digital twins will be called on to make predictions, often in extrapolatory regimes where data are limited or models are untested. An additional challenge for VVUQ is the dynamic model updating and adaptation that is key to the digital twin concept.
- Data quality, availability, and affordability are challenges. A particular challenge is the prohibitive cost of generating sufficient data for machine learning (ML) and surrogate model training.

Physical Counterpart: Foundational Research Needs and Opportunities

Digital twins rely on observation of the physical counterpart in conjunction with modeling to inform the virtual representation. In many applications, these data will be multimodal, from disparate sources, and of varying quality. While significant literature has been devoted to best practices around gathering and preparing data for use, several important gaps and opportunities are crucial for robust digital twins. Key gaps, research needs, and opportunities include the following:

- Undersampling in complex systems with large spatiotemporal variability is a significant challenge for acquiring the data needed for digital twin development. Understanding and quantifying this uncertainty is vital for assessing the reliability and limitations of the digital twin, especially in safety-critical or high-stakes applications.
- Tools are needed for data and metadata handling and management to ensure that data and metadata are gathered, recorded, stored, and processed efficiently.
- Mathematical tools are needed for assessing data quality, determining appropriate utilization of all available information, and understanding how data quality affects the performance of digital twin systems.
- Standards and governance policies are critical for data quality, accuracy, security, and integrity, and frameworks play an important role in providing standards and guidelines for data collection, management, and sharing while maintaining data security and privacy.

Physical-to-Virtual and Virtual-to-Physical Feedback Flows: Foundational Research Needs and Opportunities

In the digital twin feedback flow from physical to virtual, inverse problem methodologies and data assimilation are required to combine physical observations and virtual models in a rigorous, systematic, and scalable way. Specific challenges for digital twins such as calibration and updating on actionable time scales highlight foundational gaps in inverse problem and data assimilation theory, methodology, and computational approaches. ML and artificial intelligence (AI) have potential large roles to play in addressing these challenges, such as through the use of online learning techniques for continuously updating models using streaming data. In addition, in settings where data are limited due to data acquisition resource constraints, AI approaches such as active learning and reinforcement learning can help guide the collection of additional data most salient to the digital twin's objectives.

On the virtual-to-physical flowpath, the digital twin is used to drive changes in the physical counterpart itself, or in the observational systems associated with the physical counterpart through an automatic controller or a human.

Accordingly, the committee identified gaps associated with the use of digital twins for automated decision-making tasks, for providing decision support to a human decision-maker, and for decision tasks that are shared jointly within a human–agent team. There are additional challenges associated with the ethics and social implications of the use of digital twins in decision-making. Key gaps, research needs, and opportunities in the physical-to-virtual and virtual-to-physical feedback flows include the following:

- Methods to incorporate state-of-the-art risk metrics and characterization of extreme events in digital twin decision-making are needed.
- The assimilation of sensor data for using digital twins on actionable time scales will require advancements in data assimilation methods and tight coupling with the control or decision-support task at hand. Data assimilation techniques are needed for data from multiple sources at different scales and numerical models with different levels of uncertainty.
- Methods and tools are needed to make sensitivity information more readily available for model-centric digital twins, including automatic differentiation capabilities that will be successful for multiphysics, multiscale digital twin virtual representations, including those that couple multiple codes, each simulating different components of a complex system. Scalable and efficient optimization and uncertainty quantification methods that handle non-differentiable functions that arise with many risk metrics are also lacking.
- Scalable methods are needed for goal-oriented sensor steering and optimal experimental design that encompass the full sense–assimilate–predict– control–steer cycle while accounting for uncertainty.
- Development of implementation science around digital twins, user-centered design of digital twins, and effective human–digital twin teaming is needed.
- Research is needed on the impact of the content, context, and mode of human–digital twin interaction on the resulting decisions.

Ethics, Privacy, Data Governance, and Security

Protecting individual privacy requires proactive ethical consideration at every phase of development and within each element of the digital twin ecosystem. Moreover, the tight integration between the physical system and its virtual representation has significant cybersecurity implications, beyond what has historically been needed, that must be considered in order to effectively safeguard and scale digital twins. While security issues with digital twins share common challenges with cybersecurity issues in other settings, the close relationship between cyber and physical in digital twins could make cybersecurity more challenging. Privacy, ownership, and responsibility for data accuracy in complex, heterogeneous digital twin environments are all areas with important open questions that require attention. While the committee noted that many data ethics and governance issues fall outside the study's charge, it is important to highlight the dangers of scaling digital twins without actionable standards for appropriate use and guidelines for identifying liability in the case of misuse. Furthermore, digital twins necessitate heightened levels of security, particularly around the transmission of data and information between the physical and virtual counterparts. Especially in sensitive or high-risk settings, malicious interactions could result in security risks for the physical system. Additional safeguard design is necessary for digital twins.

TOWARD SCALABLE AND SUSTAINABLE DIGITAL TWINS

Realizing the societal benefits of digital twins will require both incremental and more dramatic research advances in cross-disciplinary approaches. In addition to bridging fundamental research challenges in statistics, mathematics, and computing, bringing complex digital twins to fruition necessitates robust and reliable yet agile and adaptable integration of all these disparate pieces.

Evolution and Sustainability of a Digital Twin

Over time, the digital twin will likely need to meet new demands, incorporate new or updated models, and obtain new data from the physical system to maintain its accuracy. Model management is key for supporting the digital twin evolution. For a digital twin to faithfully reflect temporal and spatial changes where applicable in the physical counterpart, the resulting predictions must be reproducible, incorporate improvements in the virtual representation, and be reusable in scenarios not originally envisioned. This, in turn, requires a design approach to digital twin development and evolution that is holistic, robust, and enduring, yet flexible, composable, and adaptable. Digital twins require a foundational backbone that, in whole or in part, is reusable across multiple domains, supports multiple diverse activities, and serves the needs of multiple users. Digital twins must seamlessly operate in a heterogeneous and distributed infrastructure supporting a broad spectrum of operational environments, ranging from hand-held mobile devices accessing digital twins on-the-go to large-scale, centralized high-performance computing installations. Sustaining a robust, flexible, dynamic, accessible, and secure digital twin is a key consideration for creators, funders, and the diverse community of stakeholders.

Conclusion 7-1: The notion of a digital twin has inherent value because it gives an identity to the virtual representation. This makes the virtual representation—the mathematical, statistical, and computational models of the system and its data—an asset that should receive investment and sustainment in ways that parallel investment and sustainment in the physical counterpart.

Recommendation 4: Federal agencies should each conduct an assessment for their major use cases of digital twin needs to maintain and sustain data, software, sensors, and virtual models. These assessments should drive the definition and establishment of new programs similar to the National Science Foundation's Natural Hazards Engineering Research Infrastructure and Cyberinfrastructure for Sustained Scientific

Innovation programs. These programs should target specific communities and provide support to sustain, maintain, and manage the life cycle of digital twins beyond their initial creation, recognizing that sustainability is critical to realizing the value of upstream investments in the virtual representations that underlie digital twins.

Translation and Collaborations Between Domains

There are domain-specific and even use-specific digital twin challenges, but there are also many elements that cut across domains and use cases. For digital twin virtual representations, advancing the models themselves is necessarily domain-specific, but advancing the digital twin enablers of hybrid modeling and surrogate modeling embodies shared challenges that crosscut domains. For the physical counterpart, many of the challenges around sensor technologies and data are domain-specific, but issues around handling and fusing multimodal data, enabling access to data, and advancing data curation practices embody shared challenges that crosscut domains. When it comes to the physical-to-virtual and virtual-to-physical flows, there is an opportunity to advance data assimilation, inverse methods, control, and sensor-steering methodologies that are applicable across domains, while at the same time recognizing domain-specific needs, especially as they relate to the domain-specific nature of decision-making. Finally, there is a significant opportunity to advance digital twin VVUQ methods and practices in ways that translate across domains.

As stakeholders consider architecting programs that balance these domain-specific needs with cross-domain opportunities, it is important to recognize that different domains have varying levels of maturity with respect to the different elements of the digital twin. For example, the Earth system science community is a leader in data assimilation; many fields of engineering are leaders in integrating VVUQ into simulation-based decision-making; and the medical community has a strong culture of prioritizing the role of a human decision-maker when advancing new technologies. Cross-domain interactions through the common lens of digital twins are opportunities to share, learn, and cross-fertilize.

Conclusion 7-2: As the foundations of digital twins are established, it is the ideal time to examine the architecture, interfaces, bidirectional workflows of the virtual twin with the physical counterpart, and community practices in order to make evolutionary advances that benefit all disciplinary communities.

Recommendation 5: Agencies should collaboratively and in a coordinated fashion provide cross-disciplinary workshops and venues to foster identification of those aspects of digital twin research and development that would benefit from a common approach and which specific research

topics are shared. Such activities should encompass responsible use of digital twins and should necessarily include international collaborators.

Recommendation 6: Federal agencies should identify targeted areas relevant to their individual or collective missions where collaboration with industry would advance research and translation. Initial examples might include the following:
- **Department of Defense—asset management, incorporating the processes and practices employed in the commercial aviation industry for maintenance analysis.**
- **Department of Energy—energy infrastructure security and improved (efficient and effective) emergency preparedness.**
- **National Institutes of Health—in silico drug discovery, clinical trials, preventative health care and behavior modification programs, clinical team coordination, and pandemic emergency preparedness.**
- **National Science Foundation—Directorate for Technology, Innovation and Partnerships programs.**

There is a history of both sharing and coordination of models within the international climate research community as well as a consistent commitment to data exchange that is beneficial to digital twins. While other disciplines have open-source or shared models, few support the breadth in scale and the robust integration of uncertainty quantification that are found in Earth system models and workflows. A greater level of coordination among the multidisciplinary teams of other complex systems, such as biomedical systems, would benefit maturation and cultivate the adoption of digital twins.

Conclusion 7-4: Fostering a culture of collaborative exchange of data and models that incorporate context through metadata and provenance in digital twin–relevant disciplines could accelerate progress in the development and application of digital twins.

Recommendation 7: In defining new digital twin research efforts, federal agencies should, in the context of their current and future mission priorities, (1) seed the establishment of forums to facilitate good practices for effective collaborative exchange of data and models across disciplines and domains, while addressing the growing privacy and ethics demands of digital twins; (2) foster and/or require collaborative exchange of data and models; and (3) explicitly consider the role for collaboration and coordination with international bodies.

Preparing an Interdisciplinary Workforce for Digital Twins

The successful adoption and progress of digital twins hinge on the appropriate education and training of the workforce. This educational shift requires formalizing, nurturing, and growing critical computational, mathematical, and engineering skill sets at the intersection of disciplines such as biology, chemistry, and physics. These critical skill sets include but are not limited to systems engineering, systems thinking and architecting, data analytics, ML/AI, statistical/probabilistic modeling and simulation, uncertainty quantification, computational mathematics, and decision science. These disciplines are rarely taught within the same academic curriculum.

Recommendation 8: Within the next year, federal agencies should organize workshops with participants from industry and academia to identify barriers, explore potential implementation pathways, and incentivize the creation of interdisciplinary degrees at the bachelor's, master's, and doctoral levels.

REFERENCES

AIAA (American Institute of Aeronautics and Astronautics) Digital Engineering Integration Committee. 2020. "Digital Twin: Definition & Value." AIAA and AIA Position Paper, AIAA, Reston, VA.

NRC (National Research Council). 2012. *Assessing the Reliability of Complex Models: Mathematical and Statistical Foundations of Verification, Validation, and Uncertainty Quantification*. Washington, DC: The National Academies Press.

1

Introduction

Digital twins, which are virtual representations of natural, engineered, or social systems, hold immense promise in accelerating scientific discovery and revolutionizing industries. This report aims to shed light on the key research needs to advance digital twins in several domains, and the opportunities that can be realized by bridging the gaps that currently hinder the effective implementation of digital twins in scientific research and industrial processes. This report provides practical recommendations to bring the promise of digital twins to fruition, both today and in the future.

THE SIGNIFICANCE OF DIGITAL TWINS

Digital twins are being explored and implemented in various domains as tools to allow for deeper insights into the performance, behavior, and characteristics of natural, engineered, or social systems. A digital twin can be a critical tool for decision-making that uses a synergistic combination of models and data. The bidirectional interplay between models and data endows the digital twin with a dynamic nature that goes beyond what has been traditionally possible with modeling and simulation, creating a virtual representation that evolves with the system over time. The use cases for digital twins are diverse and proliferating—including applications in biomedical research, engineering, atmospheric science, and many more—and their potential is wide-reaching.

Digital twins are emerging as enablers for significant, sustainable progress across industries. With the potential to transform traditional scientific and industrial practices and enhance operational efficiency, digital twins have captured the attention and imagination of professionals across various disciplines and sectors. By simulating real-time behavior, monitoring performance to detect anomalies and exceptional conditions, and enabling predictive insights and effective optimizations, digital twins have the capacity to revolutionize scientific research, enhance operational efficiency, optimize production strategies, reduce time-to-market, and unlock new avenues for scientific and industrial growth and innovation.

Digital twins not only offer a means to capture the knowledge and expertise of experienced professionals but also provide a platform for knowledge transfer and continuity. By creating a digital representation of assets and systems, organizations can bridge the gap between generations, ensuring that critical knowledge is preserved and accessible to future workforces and economies.

In the present landscape, "digital twin" has become a buzzword, often associated with innovation and transformation. While there is significant enthusiasm around industry developments and applications of digital twins, the focus of this report is on identifying research gaps and opportunities. The report's recommendations are particularly targeted toward what agencies and researchers can do to advance mathematical, statistical, and computational foundations of digital twins. Scientific and industrial organizations are eager to explore the possibilities offered by digital twins, but gaps and challenges often arise that impede their implementation and hinder their ability to fully deliver the promised value. Organizations eager to use digital twins do not always understand how well the digital twins match reality and whether they can be relied on for critical decisions—much of this report is aimed at elucidating the foundational mathematical, statistical, and computational research needed to bridge those gaps. Other technological complexities pose challenges as well, such as network connectivity and edge computing capabilities, data integration issues and the lack of standardized frameworks or data structures, and interoperability among various systems. Additional challenges include organizational aspects, including workforce readiness, cultural shifts, and change management required to facilitate the successful adoption and integration of digital twins. Furthermore, ensuring data security, cybersecurity, privacy, and ethical practices remains a pressing concern as organizations delve into the realm of digital twins.

COMMITTEE TASK AND SCOPE OF WORK

This study was supported by the Department of Energy (Office of Advanced Scientific Computing Research and Office of Biological and Environmental Research), the Department of Defense (Air Force Office of Scientific Research and Defense Advanced Research Projects Agency), the National Institutes of Health (National Cancer Institute, National Institute of Biomedical Imaging and Bioengineering, National Library of Medicine, and Office of Data Science Strategy), and the National Science Foundation (Directorate for Engineering and Directorate for Mathematical and Physical Sciences). In collaboration with the National

Academies of Sciences, Engineering, and Medicine, these agencies developed the study's statement of task (see Appendix A), which highlights important questions relating to the following:

- Definitions of and use cases for digital twins;
- Foundational mathematical, statistical, and computational gaps for digital twins;
- Best practices for digital twin development and use; and
- Opportunities to advance the use and practice of digital twins.

The National Academies appointed a committee of 16 members with expertise in mathematics, statistics, computer science, computational science, data science, uncertainty quantification, biomedicine, computational biology, other life sciences, engineering, atmospheric science and climate, privacy and ethics, industry, urban planning/smart cities, and defense. Committee biographies are provided in Appendix F.

The committee held several data-gathering meetings in support of this study, including three public workshops on the use of digital twins in atmospheric and climate sciences (NASEM 2023a), biomedical sciences (NASEM 2023b), and engineering (NASEM 2023c).

REPORT STRUCTURE

This report was written with the intention of informing the scientific and research community, academia, pertinent government agencies, digital twin practitioners, and those in relevant industries about open needs and foundational gaps to overcome to advance digital twins. While the range of challenges and open questions around digital twins is broad, it should be noted that the focus of this report is on foundational gaps. The report begins by defining a digital twin, outlining its elements and overarching themes, and articulating the need for an integrated research agenda in Chapter 2. The next four chapters expound on the four major elements of a digital twin as defined by the committee: the virtual representation, the physical counterpart, the feedback flow from the physical to the virtual, and the feedback flow from the virtual to the physical. In Chapter 3, fitness for purpose, modeling challenges, and integration of digital twin components for the virtual representation are discussed. Chapter 4 explores the needs and opportunities around data acquisition and data integration in preparation for inverse problem and data assimilation tasks, which are discussed in Chapter 5. Automated decision-making and human–digital twin interactions, as well as the ethical implications of making decisions using a digital twin or its outputs, are addressed in Chapter 6. Chapter 7 looks at some of the broader gaps and needs to be addressed in order to scale and sustain digital twins, including cross-community efforts and workforce challenges. Finally, Chapter 8 aggregates all of the findings, conclusions, gaps, and recommendations placed throughout the report.

REFERENCES

NASEM (National Academies of Sciences, Engineering, and Medicine). 2023a. *Opportunities and Challenges for Digital Twins in Atmospheric and Climate Sciences: Proceedings of a Workshop—in Brief.* Washington, DC: The National Academies Press.

NASEM. 2023b. *Opportunities and Challenges for Digital Twins in Biomedical Research: Proceedings of a Workshop—in Brief.* Washington, DC: The National Academies Press.

NASEM. 2023c. *Opportunities and Challenges for Digital Twins in Engineering: Proceedings of a Workshop—in Brief.* Washington, DC: The National Academies Press.

2

The Digital Twin Landscape

This chapter lays the foundation for an understanding of the landscape of digital twins and the need for an integrated research agenda. The chapter begins by defining a digital twin. It then articulates the elements of the digital twin ecosystem, discussing how a digital twin is more than just a simulation and emphasizing the bidirectional interplay between a virtual representation and its physical counterpart. The chapter discusses the critical role of verification, validation, and uncertainty quantification (VVUQ) in digital twins, as well as the importance of ethics, privacy, data governance, and security. The chapter concludes with a brief assessment of the state of the art and articulates the importance of an integrated research agenda to realize the potential of digital twins across science, technology, and society.

DEFINITIONS

Noting that the scope of this study is on identifying foundational research gaps and opportunities for digital twins, it is important to have a shared understanding of the definition of a digital twin. For the purposes of this report, the committee uses the following definition of a digital twin:

> A digital twin is a set of virtual information constructs that mimics the structure, context, and behavior of a natural, engineered, or social system (or system-of-systems), is dynamically updated with data from its physical twin, has a predictive capability, and informs decisions that realize value. The bidirectional interaction between the virtual and the physical is central to the digital twin.

This definition is based heavily on a definition published in 2020 by the American Institute of Aeronautics and Astronautics (AIAA) Digital Engineering Integration Committee (2020). The study committee's definition modifies the AIAA definition to better align with domains beyond aerospace engineering. In place of the term "asset," the committee refers to "a natural, engineered, or social system (or system-of-systems)" to describe digital twins of physical systems in the broadest sense possible, including the engineered world, natural phenomena, biological entities, and social systems. The term "system-of-systems" acknowledges that many digital twin use cases involve virtual representations of complex systems that are themselves a collection of multiple coupled systems. This definition also introduces the phrase "has a predictive capability" to emphasize the important point that a digital twin must be able to issue predictions beyond the available data in order to drive decisions that realize value. Finally, the committee's definition adds the sentence "The bidirectional interaction between the virtual and the physical is central to the digital twin." As described below, the bidirectional interaction comprises feedback flows of information from the physical system to the virtual representation and from the virtual back to the physical system to enable decision-making, either automatic or with a human- or humans-in-the-loop. Although the importance of the bidirectional interaction is implicit in the earlier part of the definition, our committee's information gathering revealed the importance of explicitly emphasizing this aspect (Ghattas 2023; Girolami 2022; Wells 2022).

While it is important to have a shared understanding of the definition of a digital twin, it is also important to recognize that the broad nature of the digital twin concept will lead to differences in digital twin elements across different domains, and even in different use cases within a particular domain. Thus, while the committee adopts this definition for the purposes of this report, it recognizes the value in alternate definitions in other settings.

Digital Twin Origins

While the concept itself is older, the term "digital twin" emerged around 2010 during technical roadmapping efforts at NASA co-led by John Vickers. The term "digital twin" was defined in published NASA reports by Piascik et al. (2012) and Shafto et al. (2012), and in a follow-on paper by Glaessgen and Stargel (2012):[1]

> A digital twin is an integrated multiphysics, multi-scale, probabilistic simulation of a vehicle or system that uses the best available physical models, sensor updates, fleet history, etc., to mirror the life of its flying twin. The digital twin is ultra-realistic and may consider one or more important and interdependent vehicle systems, including propulsion and energy storage, life support, avionics, thermal protection, etc. (Shafto et al. 2012)

---

[1] This paragraph was changed after the release of the report to accurately reflect the emergence of the term "digital twin." These NASA reports were released publicly in 2010 but have 2012 official publication dates.

This definition and notion are built on earlier work by Grieves (2005a,b) in product life-cycle[2] management. A closely related concept is that of Dynamic Data Driven Application Systems (DDDAS) (Darema 2004). Some of the early published DDDAS work has all the elements of a digital twin, including the physical, the virtual, and the two-way interaction via a feedback loop. Many of the notions underlying digital twins also have a long history in other fields, such as model predictive control, which similarly combines models and data in a bidirectional feedback loop (Rawlings et al. 2017), and data assimilation, which has long been used in the field of weather forecasting to combine multiple sources of data with numerical models (Reichle 2008).

Much of the early work and development of digital twins was carried out in the field of aerospace engineering, particularly in the use of digital twins for structural health monitoring and predictive maintenance of airframes and aircraft engines (Tuegel et al. 2011). Today, interest in and development of digital twins has expanded well beyond aerospace engineering to include many different application areas across science, technology, and society. With that expansion has come a broadening in the views of what constitutes a digital twin along with differing specific digital twin definitions within different application contexts. During information-gathering sessions, the committee heard multiple different definitions of digital twins. The various definitions have some common elements, but even these common elements are not necessarily aligned across communities, reflecting the different nature of digital twins in different application settings. The committee also heard from multiple briefers that the "Digital Twin has no common agreed definition" (Girolami 2022; NASEM 2023a,b,c).

ELEMENTS OF THE DIGITAL TWIN ECOSYSTEM

A Digital Twin Is More Than Just Simulation and Modeling

The notion of a digital twin builds on a long history of modeling and simulation of complex systems but goes beyond simulation to include tighter integration between models, observational data, and decisions. The dynamic, bidirectional interaction between the physical and the virtual enables the digital twin to be tailored to a particular physical counterpart and to evolve as the physical counterpart evolves. This, in turn, enables dynamic data-driven decision-making.

Finding 2-1: A digital twin is more than just simulation and modeling.

Conclusion 2-1: The key elements that comprise a digital twin include (1) modeling and simulation to create a virtual representation of a physical

---

[2] For the purposes of this report, the committee defines *life cycle* as the "overall process of developing, implementing, and retiring ... systems through a multistep process from initiation, analysis, design, implementation, and maintenance to disposal" as defined in NIST (2009).

counterpart, and (2) a bidirectional interaction between the virtual and the physical. This bidirectional interaction forms a feedback loop that comprises dynamic data-driven model updating (e.g., sensor fusion, inversion, data assimilation) and optimal decision-making (e.g., control, sensor steering).

These elements are depicted abstractly in Figure 2-1 and with examples in Box 2-1. More details are provided in the following subsections.

The Physical Counterpart and Its Virtual Representation

There are numerous and diverse examples of physical counterparts for which digital twins are recognized as bringing high potential value, including aircraft, body organs, cancer tumors, cities, civil infrastructure, coastal areas, farms, forests, global atmosphere, hospital operations, ice sheets, nuclear reactors, patients, and many more. These examples illustrate the broad potential scope of a digital twin, which may bring value at multiple levels of subsystem and system modeling. For example, digital twins at the levels of a cancer tumor, a body organ, and a patient all have utility and highlight the potential trade-offs in digital twin scope versus complexity. Essential to being able to create digital twins is the ability to acquire data from the physical counterpart. These data may be acquired from onboard or in situ sensors, remote sensing, automated and visual inspections, operational logs, imaging, and more. The committee considers these sensing and observational systems to be a part of the physical counterpart in its representation of the digital twin ecosystem.

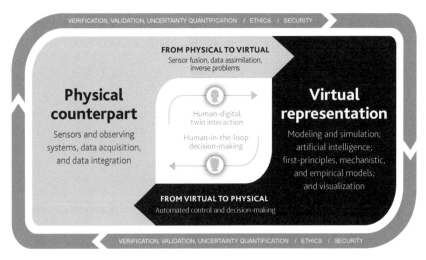

FIGURE 2-1 Elements of the digital twin ecosystem.
NOTES: Information flows bidirectionally between the virtual representation and physical counterpart. These information flows may be through automated processes, human-driven processes, or a combination of the two.

BOX 2-1
Digital Twin Examples

Digital Twin of a Cancer Patient (Figure 2-1-1)

The virtual representation of a cancer patient might comprise mechanistic models in the form of nonlinear partial differential equations describing temporal and spatial characteristics of tumor growth, with a state variable that represents spatiotemporal tumor cell density and/or heterogeneity. These models are characterized by parameters that represent the specific patient's anatomy, morphology, and constitutive properties such as the tumor cell proliferation rate and tissue carrying capacity; parameters that describe the initial tumor location, geometry, and burden; and parameters that describe the specific patient's response to treatments such as radiotherapy, chemotherapy, and immunotherapy. Quantities of interest might include computational estimates of patient characteristics, such as tumor cell count, time to progression, and toxicity. Decision tasks might include personalized therapy control decisions, such as the dose and schedule of delivery of therapeutics over time, and data collection decisions, such as the frequency of serial imaging studies, blood tests, and other clinical assessments. These decisions can be automated as part of the digital twin or made by a human informed by the digital twin's output.[a]

Digital Twin of an Aircraft Engine

The virtual representation of an aircraft engine might comprise machine learning (ML) models trained on a large database of sensor data and flight logs collected across a fleet of engines. These models are characterized by parameters that represent the operating conditions seen by this particular engine and numerical model parameters that represent the hyperparameters of the ML models. Quantities of interest might include computational estimates of possible blade material degradation. Decision tasks might include actions related to what maintenance to perform and when, as well as decisions related to performing additional inspections; these actions can be taken by a human informed by the digital twin's output, or they can be taken automatically by the digital twin.[b] For instance, the digital twin could be leveraged for optimizing fuel efficiency in real time, simulating emergency response scenarios for enhanced pilot training, predicting parts that may soon need replacement for efficient inventory management, ensuring regulatory compliance on environmental and safety fronts, conducting cost–benefit analyses of various maintenance strategies, controlling noise pollution levels, and even assessing and planning for carbon emission reduction. By incorporating these additional decision-making tasks, the digital twin can contribute more comprehensively to the aircraft engine's operational efficiency, safety protocols, and compliance with environmental standards, thus amplifying its utility beyond merely informing maintenance schedules.

continued

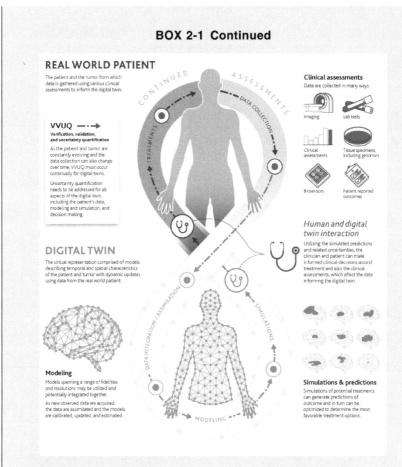

FIGURE 2-1-1 Example of a digital twin of a cancer patient and tumor.

Digital Twin of an Earth System

The virtual representation of an Earth system might comprise a collection of high-fidelity, high-resolution physics models and associated surrogate models, collectively representing coupled atmospheric, oceanic, terrestrial, and cryospheric

---

The virtual representation of the physical counterpart comprises a computational model or set of coupled models. These models are typically computational representations of first-principles, mechanistic, and/or empirical models, which take on a range of mathematical forms, including dynamical systems, differential equations, and statistical models (including machine learning [ML] models). The set of models comprising the virtual representation of a digital twin of a complex system will span multiple disciplines and multiple temporal and spatial scales.

physics. These models solve for state variables such as pressure, temperature, density, and salinity. These models are characterized by parameters that represent physical properties such as terrain geometry, fluid constitutive properties, boundary conditions, initial conditions, and anthropogenic source terms, as well as numerical model parameters that represent, for example, turbulence model closures and ML model hyperparameters. Quantities of interest might include projections of future global mean temperature or statistics of extreme precipitation events. Decision tasks might include actions related to policy-making, energy system design, deployment of new observing systems, and emergency preparedness for extreme weather events. These decisions may be made automatically as part of the digital twin or made by a human informed by the digital twin's output.[c]

Digital Twin of a Manufacturing Process

Manufacturing environments afford many opportunities for digital twins. Consider a manufacturing system potentially comprising equipment, human workers, various stations and assembly lines, processes, and the materials that flow through the system. The virtual representation of a manufacturing process might include visually-, principles-, data-, and/or geometry-driven models which are parameterized by data such as process monitoring data (both real-time/near real-time and historical), production data, system layout, and equipment status and maintenance records.[d] These data may span much of the process life cycle. Of course, these components will be tailored to the specific process and should be fit-for-purpose. Decision tasks might include operational decisions and process control, for instance. These decisions may be made automatically as part of the digital twin or made by a human informed by the digital twin's output.

---

[a] National Academies of Sciences, Engineering, and Medicine, 2023, *Opportunities and Challenges for Digital Twins in Biomedical Research: Proceedings of a Workshop—in Brief*, Washington, DC: The National Academies Press.
[b] National Academies of Sciences, Engineering, and Medicine, 2023, *Opportunities and Challenges for Digital Twins in Engineering: Proceedings of a Workshop—in Brief*, Washington, DC: The National Academies Press.
[c] National Academies of Sciences, Engineering, and Medicine, 2023, *Opportunities and Challenges for Digital Twins in Atmospheric and Climate Sciences: Proceedings of a Workshop—in Brief*, Washington, DC: The National Academies Press.
[d] H. Latif, G. Shao, and B. Starly, 2020, "A Case Study of Digital Twin for a Manufacturing Process Involving Human Interactions," *Proceedings of 2020 Winter Simulation Conference*, https://tsapps.nist.gov/publication/get_pdf.cfm?pub_id=930232.

Digital twin examples in the literature employ models that span a range of fidelities and resolutions, from high-resolution, high-fidelity replicas to simplified surrogate models.

Another part of the digital twin virtual representation is the definition of parameters, states, and quantities of interest. The computational models are characterized by *parameters* that are the virtual representation of attributes such as geometry and constitutive properties of the physical counterpart, boundary conditions, initial conditions, external factors that influence the physical counterpart, and transfer coefficients between resolved processes and parameterized unresolved processes. Sometimes these parameters will be known, while in other cases they must be estimated from data. Some types of models may also be characterized by parameters and hyperparameters that represent numerical approximations within a model, such as Gaussian process correlation lengths, regularization hyperparameters, and neural network training weights. The committee refers to this latter class of parameters as *numerical model parameters* to distinguish them from the parameters that represent attributes of the physical system. The committee uses the term *state* to denote the solved-for quantities in a model that takes the form of a dynamical system or system of differential equations. However, the committee notes that in many cases, the distinction between parameter and state can become blurred—when a digital twin couples multiple models across different disciplines, the state of one model may be a parameter in another model. Furthermore, the committee notes that many digital twin use cases explicitly target situations where parameters are dynamically changing, requiring dynamic estimation and updating of parameters, akin to state estimation in classical settings. Lastly, the committee denotes *quantities of interest* as the metrics that are of particular relevance to digital twin predictions and decisions. These quantities of interest are typically functions of parameters and states. The quantities of interest may themselves vary in definition as a particular digital twin is used in different decision-making scenarios over time.

An important theme that runs throughout this report is the notion that the virtual representation be *fit for purpose*, meaning that the virtual representation—model types, fidelity, resolution, parameterization, and quantities of interest—be chosen, and in many cases dynamically adapted, to fit the particular decision task and computational constraints at hand. Another important theme that runs throughout this report is the critical need for uncertainty quantification to be an integral part of digital twin formulations. If this need is addressed by, for example, the use of Bayesian formulations, then the formulation of the virtual representation must also define prior information for parameters, numerical model parameters, and states.

Bidirectional Feedback Flow Between Physical and Virtual

The bidirectional interaction between the virtual representation and the physical counterpart forms an integral part of the digital twin. This interaction is sometimes characterized as a feedback loop, where data from the physical counterpart are used to update the virtual models, and, in turn, the virtual models are used to drive changes in the physical system. This feedback loop may occur in real time, such as for dynamic control of an autonomous vehicle or a wind farm, or it may occur on slower time scales, such as post-flight updating of a digital twin for aircraft engine predictive maintenance or post-imaging updating of a digital twin and subsequent treatment planning for a cancer patient.

On the physical-to-virtual flowpath, digital twin tasks include sensor data fusion, model calibration, dynamic model updating, and estimation of parameters and states that are not directly observable. These calibration, updating, and estimation tasks are typically posed mathematically as data assimilation and inverse problems, which can take the form of parameter estimation (both static and dynamic), state estimation, regression, classification, and detection.

On the virtual-to-physical flowpath, the digital twin is used to drive changes in the physical counterpart itself or in the sensor and observing systems associated with the physical counterpart. This flowpath may be fully automated, where the digital twin interacts directly with the physical system. Examples of automated decision-making tasks include automated control, scheduling, recommendation, and sensor steering. In many cases, these tasks relate to automatic feedback control, which is already in widespread use across many engineering systems. Concrete examples of potential digital twin automated decision-making tasks are given in the illustrative examples in Box 2-1. The virtual-to-physical flowpath may also include a human in the digital twin feedback loop. A human may play the key decision-making role, in which case the digital twin provides decision support, or decision-making may be shared jointly between the digital twin and a human as a human–agent team. Human–digital twin interaction may also take the form of the human playing a crucial role in designing, managing, and operating elements of the digital twin, including selecting sensors and data sources, managing the models underlying the virtual representation, and implementing algorithms and analytics tools. User-centered design is central to extracting value from the digital twin.

Verification, Validation, and Uncertainty Quantification

VVUQ is essential for the responsible development, implementation, monitoring, and sustainability of digital twins. Since the precise definitions can differ among subject-matter areas, the committee adopts the definition of VVUQ used in the National Research Council report *Assessing the Reliability of Complex Models* (NRC 2012) for this report:

- *Verification* is "the process of determining whether a computer program ('code') correctly solves the equations of the mathematical model. This includes code verification (determining whether the code correctly implements the intended algorithms) and solution verification (determining the accuracy with which the algorithms solve the mathematical model's equations for specified quantities of interest)."
- *Validation* is "the process of determining the degree to which a model is an accurate representation of the real world from the perspective of the intended uses of the model (taken from AIAA [Computational Fluid Dynamics Committee], 1998)."

- *Uncertainty quantification* is "the process of quantifying uncertainties associated with model calculations of true, physical quantities of interest, with the goals of accounting for all sources of uncertainty and quantifying the contributions of specific sources to the overall uncertainty."

Each of the VVUQ tasks plays important roles for digital twins. There are, however, key differences. The challenges lie in the features that set digital twins apart from traditional modeling and simulation, with the most important difference being the bidirectional feedback loop between the virtual and the physical. Evolution of the physical counterpart in real-world use conditions, changes in data collection hardware and software, noisiness of data, addition and deletion of data sources, changes in the distribution of the data shared with the virtual twin, changes in the prediction and/or decision tasks posed to the digital twin, and evolution of the digital twin virtual models all have consequences for VVUQ. Significant challenges remain for VVUQ of stochastic and adaptive systems; due to their dynamic nature, digital twins inherit these challenges.

Traditionally, a computational model may be verified for sets of inputs at the code verification stage and for scenarios at the solution verification stage. While many of the elements are shared with VVUQ for computational models (NRC 2012), for digital twins, one anticipates, over time, upgrades to data collection technology (e.g., sensors). This may mean changes in the quality of data being collected, more and cheaper data capture hardware with potentially lower quality of information, different data sources, or changes in data structures. Additionally, the physical counterpart's state will undergo continual evolution. With such changes comes the need to revisit some or all aspects of verification. Furthermore, as the physical twin evolves over its lifetime, it is possible to enter system states that are far from the solution scenarios that were envisioned at initial verification. Indeed, major changes made to the physical twin may require that the virtual representation be substantially redefined and re-implemented.

As with verification, validation is more complicated in the context of a digital twin. The output of a digital twin needs to include the confidence level in its prediction. Changes in the state of the physical counterpart, data collection and structures, and the computational models can each impact the validation assessment and may require continual validation. The bidirectional interplay between the physical and the virtual means the predictive model is periodically, or even continuously, updated. For continual VVUQ, automated VVUQ methods may yield operational efficiencies. These updates must be factored into digital twin validation processes.

Uncertainty quantification is essential to making informed decisions and to promoting the necessary transparency for a digital twin to build trust with decision support. Uncertainty quantification is also essential for fitness-for-purpose considerations. There are many potential sources of uncertainty in a digital twin. These include those arising from modeling uncertainties (Chapter 3), measurement and other data uncertainties (Chapter 4), the processes of data assimilation and model calibration (Chapter 5), and decision-making (Chapter 6). Particularly unique to digital twins is inclusion of uncertainties due to integration of multiple modalities of data and models, and bidirectional and sometimes real-time interaction between the virtual representation, the physical counterpart, and the possible human-in-the-loop interactions. These interactions and integration may even lead to new instabilities that emerge due to the nonlinear coupling among different elements of the digital twin.

Given the interconnectedness of different systems and stakeholders across the digital twin ecosystem, it is imperative to outline the VVUQ pipeline and highlight potential sources of information breakdown and model collapse. It is important to recognize that VVUQ must play a role in all elements of the digital twin ecosystem. In the digital twin virtual representation, verification plays a key role in building trust that the mathematical models used for simulation of the physical counterpart have been sufficiently implemented. In cases that employ surrogate models, uncertainty quantification gives measures of the quality of prediction that the surrogate model provides. Field observations, for example, can be used to estimate uncertainties and parameters that govern the virtual representation (a type of inverse problem) as a step toward model validation, followed by the assessment of predictions. As information is passed from the physical counterpart to the virtual representation, new data can be used to update estimates and predictions with uncertainty quantification that can be used for decisions. These include challenges arising from model discrepancy, unresolved scales, surrogate modeling, and the need to issue predictions in extrapolatory regimes (Chapter 3).

When constructing digital twins, there are often many sources of data (e.g., data arising from sensors or simulations), and consequently, there can be many sources of uncertainty. Despite the abundance of data, there are nonetheless limitations to the ability to reduce uncertainty. Computational models may inherently contain unresolvable model form errors or discrepancies. Additionally, measurement errors in sensors are typically unavoidable. Whether adopting a data-centric or model-centric view, it is important to assess carefully which parts of the digital twin model can be informed by data and simulations and which cannot in order to prevent overfitting and to provide a full accounting of uncertainty.

The VVUQ contribution does not stop with the virtual representation. Monitoring the uncertainties associated with the physical counterpart and incorporating changes to, for example, sensors or data collection equipment are part of ensuring data quality passed to the virtual counterpart. Data quality improvements may be prioritized based on the relative impacts of parameter uncertainties on the resulting model uncertainties. Data quality challenges arise from measurement, undersampling, and other data uncertainties (Chapter 4). Data quality is especially pertinent when ML models are used. Research into methods for identifying and mitigating the impact of noisy or incomplete data is needed. VVUQ can also play a role in understanding the impact of mechanisms used to pass information between the physical and virtual, and vice versa. These include challenges arising from parameter uncertainty and ill-posed or indeterminate inverse problems (Chapter 5). Additionally, the uncertainty introduced by the inclusion of the human-in-the-loop should be measured and quantified in some settings. The human-in-the-loop as part of the VVUQ pipeline can be a critical source of variability that also has to be taken into consideration (Chapter 6). This can be particularly important in making predictions where different decision makers are involved.

A digital twin without serious considerations of VVUQ is not trustworthy. However, a rigorous VVUQ approach across all elements of the digital twin may be difficult to achieve. Digital twins may represent systems-of-systems with multiscale, multiphysics, and multi-code components. VVUQ methods, and methods supporting digital twins broadly, will need to be adaptable and scalable as digital twins increase in complexity. Finally, the choice of performance metrics for VVUQ will depend on the use case. Such metrics might include average case prediction error (e.g., mean square prediction error), predictive variance, worse case prediction error, or risk-based assessments.

While this section has not provided an exhaustive list of VVUQ contributions to the digital twin ecosystem, it does serve to highlight that VVUQ plays a critical role in all aspects. Box 2-2 highlights the Department of Energy Predictive Science Academic Alliance Program as an exemplar model of interdisciplinary research that promotes VVUQ.

Conclusion 2-2: Digital twins require VVUQ to be a continual process that must adapt to changes in the physical counterpart, digital twin virtual models, data, and the prediction/decision task at hand. A gap exists between the class of problems that has been considered in traditional modeling and simulation settings and the VVUQ problems that will arise for digital twins.

The importance of a rigorous VVUQ process for a potentially powerful tool such as a digital twin cannot be overstated. Consider the growing concern over the dangers of artificial intelligence (AI),[3] with warnings even extending to the "risk of human extinction" (Center for A.I. Safety 2023; Roose 2023). Generative AI models such as ChatGPT are being widely deployed, despite open questions about their reliability, robustness, and accuracy. There has long been a healthy skepticism about the use of predictive simulations in critical decision-making. Over time, use-driven research and development in VVUQ has provided a robust framework to foster confidence and establish boundaries for use of simulations that draw from new and ongoing computational science research (Hendrickson et al. 2020). As a result of continued advances in VVUQ, many of the ingredients

---

[3] This report went through the review process prior to the October 30, 2023, release of the Biden-Harris administration Executive Order on the responsible development of artificial intelligence. While much of the discussion is relevant to this report, the committee did not have an opportunity to review and comment on the Executive Order as part of this study.

> **BOX 2-2**
> **Department of Energy Predictive Science Academic Alliance Program: Interdisciplinary Research Promoting Verification, Validation, and Uncertainty Quantification**
>
> For more than two decades, the Department of Energy (DOE) National Nuclear Security Administration's (NNSA's) Advanced Simulation and Computing Program (ASC) has proven an exemplary model for promoting interdisciplinary research in computational science in U.S. research universities, which deserves emulation by other federal agencies.
>
> ASC has established a strong portfolio of strategic alliances with leading U.S. academic institutions. The program was established in 1997 to engage the U.S. academic community in advancing science-based modeling and simulation technologies.[a] At the core of each university center is an overarching complex multiphysics problem that requires innovations in programming and runtime environments, physical models and algorithms, data analysis at scale, and uncertainty analysis. This overarching problem (proposed independently by each center) has served as a most effective catalyst to promote interdisciplinary cooperation among multiple departments (e.g., mathematics, computer science, and engineering). In 2008, a new phase of the ASC alliance program, the Predictive Science Academic Alliance Program (PSAAP),[b] added an emphasis on verification, validation, and uncertainty quantification (VVUQ). PSAAP has profoundly affected university cultures and curricula in computational science by infusing VVUQ; scalable computing; programming paradigms on heterogeneous computer systems; multiscale, multiphysics, and multi-code integration science; etc. To facilitate the research agendas of the centers, DOE/NNSA provides significant cycles on the most powerful unclassified computing systems.
>
> An important aspect of the management of PSAAP involves active interactions with the scientists at the NNSA laboratories through biannual rigorous technical reviews that focus on the technical progress of the centers and provide recommendations to help them meet their goals and milestones. Another important aspect is required graduate student internships at the NNSA laboratories.
>
> ---
> [a] Lawrence Livermore National Laboratory, n.d., "Archive: A Brief History of ASC and PSAAP," https://psaap.llnl.gov/archive, accessed August 11, 2023.
> [b] National Nuclear Security Administration, n.d., "Predictive Science Academic Alliance Program," https://www.nnsa-ap.us/Programs/Predictive-Science-Academic-Alliance-Program, accessed August 8, 2023.

of AI methods—statistical modeling, surrogate modeling, inverse problems, data assimilation, optimal control—have long been used in engineering and scientific applications with acceptable levels of risk. One wonders: Is it the methods themselves that pose a risk to the human enterprise, or is it the way in which they are deployed without due attention to VVUQ and certification? When it comes to digital twins and their deployment in critical engineering and scientific applications, humanity cannot afford the cavalier attitude that pervades other applications of AI. It is critical that VVUQ be deeply embedded in the design, creation, and deployment of digital twins—while recognizing that doing so will almost certainly slow progress.

Conclusion 2-3: Despite the growing use of artificial intelligence, machine learning, and empirical modeling in engineering and scientific applications, there is a lack of standards in reporting VVUQ as well as a lack of consideration of confidence in modeling outputs.

Conclusion 2-4: Methods for ensuring continual VVUQ and monitoring of digital twins are required to establish trustworthiness. It is critical that VVUQ be deeply embedded in the design, creation, and deployment of digital twins. In future digital twin research developments, VVUQ should play a core role and tight integration should be emphasized. Particular areas of research need include continual verification, continual validation, VVUQ in extrapolatory conditions, and scalable algorithms for complex multiscale, multiphysics, and multi-code digital twin software efforts.

Finding 2-2: The Department of Energy Predictive Science Academic Alliance Program has proven an exemplary model for promoting interdisciplinary research in computational science in U.S. research universities and has profoundly affected university cultures and curricula in computational science in the way that VVUQ is infused with scalable computing, programming paradigms on heterogeneous computer systems, and multiphysics and multi-code integration science.

Ethics, Privacy, and Data Governance

Protecting individual privacy requires proactive ethical consideration at every phase of development and within each element of the digital twin ecosystem. When data are collected, used, or traded, the protection of the individual's identity is paramount. Despite the rampant collection of data in today's information landscape, questions remain around preserving individual privacy. Current privacy-preserving methods, such as differential privacy or the use of synthetic data, are gaining traction but have limitations in many settings (e.g., reduced accuracy in data-scarce settings). Additionally, user data are frequently repurposed or sold. During the atmospheric and climate sciences digital twin workshop, for instance, speakers pointed out that the buying and selling of individual location data is a particularly significant challenge that deserves greater attention (NASEM 2023a).

Moreover, digital twins are enabled through the development and deployment of myriad complex algorithms. In both the biomedical workshop and atmospheric and climate sciences workshop on digital twins, speakers warned of the bias inherent in algorithms due to missing data as a result of historical and systemic biases (NASEM 2023a,b).

Collecting and using data in a way that is socially responsible, maintaining the privacy of individuals, and reducing bias in algorithms through inclusive and representative data gathering are all critical to the development of digital twins. However, these priorities are challenges for the research and professional communities at large and are not unique to digital twins. Below, the committee identifies some novel challenges that arise in the context of digital twins.

By virtue of the personalized nature of a digital twin (i.e., the digital twin's specificity to a unique asset, human, or system), the virtual construct aggregates sensitive data, potentially identifiable or re-identifiable, and models that offer tailored insights about the physical counterpart. Speakers in the biomedical digital twin workshop remarked that a digital twin in a medical setting might include a patient's entire health history and that a digital twin "will never be completely de-identifiable" (NASEM 2023b). As a repository of sensitive information, digital twins are vulnerable to data breaches, both accidental and malicious.

Speakers in both the biomedical workshop and the atmospheric and climate sciences workshop urged digital twin users and developers to enforce fitness for purpose and consider how the data are used. In a briefing to the committee, Dr. Lea Shanley repeated these concerns and stressed that the term "open data" does not mean unconditional use (Shanley 2023). During the atmospheric and climate sciences workshop, Dr. Michael Goodchild warned that "repurposing" data is a serious challenge that must be addressed (Goodchild 2023). Moreover, speakers highlighted the need for transparency surrounding individual data. As part of the final panel discussion in the biomedical workshop, Dr. Mangravite noted that once guidelines around data control are established, further work is needed to determine acceptable data access (Mangravite 2023).

The real-time data collection that may occur as part of some digital twins raises important questions around governance (NASEM 2023b). Dr. Shanley pointed out that using complex data sets that combine personal, public, and commercial data is fraught with legal and governance questions around ownership and responsibility (Shanley 2023). Understanding who is accountable for data accuracy is nontrivial and will require new legal frameworks.

Privacy, ownership, and responsibility for data accuracy in complex, heterogeneous digital twin environments are all areas with important open questions that require attention. The committee deemed governance to fall outside this study's focus on foundational mathematical, statistical, and computational gaps. However, the committee would be remiss if it did not point out the dangers of scaling (or even developing) digital twins without clear and actionable standards for appropriate use and guidelines for identifying liability in the case of accidental or intentional misuse of a digital twin or its elements, as well as mechanisms for enforcing appropriate use.

Finding 2-3: Protecting privacy and determining data ownership and liability in complex, heterogeneous digital twin environments are unresolved challenges that pose critical barriers to the responsible development and scaling of digital twins.

Finally, making decisions based on information obtained from a digital twin raises additional ethical concerns. These challenges are discussed further in the context of automated and human-in-the-loop decision-making as part of Chapter 6.

Security

Characteristic of digital twins is the tight integration between the physical system and its virtual representation. This integration has several cybersecurity implications that must be considered, beyond what has historically been needed, in order to effectively safeguard and scale digital twins.

To maximize efficacy and utility of the digital twin, the physical counterpart must share as much of its data on a meaningful time scale as possible. The need to capture and transmit detailed and time-critical information exposes the physical system to considerably more risks. Examples include physical manipulation while feeding the digital twin fake data, misleading the operator of the physical counterpart, and intercepting data traffic to capture detailed data on the physical system.

As shown in Figure 2-1, feedback is integral to the digital twin paradigm. The close integration of physical and digital systems exposes an additional attack surface for the physical system. A malicious actor can inject an attack into the feedback loop (e.g., spoofing as the digital twin) and influence the physical system in a harmful manner.

An additional novel area of security consideration for digital twins arises from the vision of an ideal future where digital twins scale easily and effortlessly. Imagine the scenario where the digital twin is exposed to the broader community (either by design or inadvertently). Since the digital twin represents true physical traits and behaviors of its counterpart, malicious interactions with the digital twin could lead to security risks for the physical system. For example, consider the digital twin of an aircraft system; a malicious actor could manipulate the digital twin to observe vulnerable traits or behaviors of the physical system (e.g., because such traits or behaviors can be inferred from certain simulations, or changes in simulation parameters). These vulnerabilities may be unknown to the system operator. A malicious actor could also interrogate the digital twin to glean intellectual property data such as designs and system parameters. Therefore, scaling digital twins must take into consideration a balance of scalability and information sharing.

DIGITAL TWIN STATE OF THE ART AND DOMAIN-SPECIFIC CHALLENGES

During information-gathering sessions, the committee heard multiple examples of potential use cases for digital twins and some practical examples of digital twins being deployed. Use cases and practical examples arising in the domains of engineering, biomedical sciences, and atmospheric and climate sciences are summarized in the three Proceedings of a Workshop—in Brief (NASEM 2023a,b,c). Practical examples of digital twins for single assets and systems of assets are also given in a recent white paper from The Alan Turing Institute (Bennett et al. 2023). Digital twins can be seen as "innovation enablers" that are redefining engineering processes and multiplying capabilities to drive innovation across industries, businesses, and governments. This level of innovation is facilitated by a digital twin's ability to integrate a product's entire life cycle with performance data and to employ a continuous loop of optimization. Ultimately, digital twins could reduce risk, accelerate time from design to production, and improve decision-making as well as connect real-time data with virtual representations for remote monitoring, predictive capabilities, collaboration among stakeholders, and multiple training opportunities (Bochenek 2023).

While the exploration and use of digital twins is growing across domains, many state-of-the-art digital twins are largely the result of custom implementations that require considerable deployment resources and a high level of expertise (Niederer et al. 2021). Many of the exemplar use cases are limited to specific applications, using bespoke methods and technologies that are not widely applicable across other problem spaces. In part as a result of the bespoke nature of many digital twin implementations, the relative maturity of digital twins varies significantly across problem spaces. This section explores some current efforts under way in addition to domain-specific needs and opportunities within aerospace and defense applications; atmospheric, climate, and sustainability sciences; and biomedical applications.

Digital Twin Examples, Needs, and Opportunities for Aerospace and Defense Applications

There are many exciting and promising directions for digital twins in aerospace and defense applications. These directions are discussed in greater detail in *Opportunities and Challenges for Digital Twins in Engineering: Proceedings of a Workshop—in Brief* in Appendix E (NASEM 2023c); the following section outlines overarching themes from the workshop. The U.S. Air Force Research Laboratory Airframe Digital Twin program focuses on better maintaining the structural integrity of military aircraft. The initial goal of the program was to use digital twins to balance the need to avoid the unacceptable risk of catastrophic failure with the need to reduce the amount of downtime for maintenance and prevent complicated and expensive maintenance. The use of data-informed simulations provides timely and actionable information to operators about what maintenance to perform and when. Operators can then plan for downtime, and maintainers can prepare to execute maintenance packages tailored for each physical twin and the corresponding asset (Kobryn 2023). The Department of Defense (DoD) could benefit from the broader use of digital twins in asset management, incorporating the processes and practices employed in the commercial aviation industry for maintenance analysis (Gahn 2023). Opportunities for digital twins include enhanced asset reliability, planned maintenance, reduced maintenance and inspection burden, and improved efficiency (Deshmukh 2023).

Significant gaps remain before the Airframe Digital Twin can be adopted by DoD. Connecting the simulations across length scales and physical phenomena is key, as is integrating probabilistic analysis. There is value in advancing optimal experimental design, active learning, optimal sensor placement, and dynamic sensor scheduling. These are significant areas of opportunity for development of digital twins across DoD applications. For example, by using simulations to determine which test conditions to run and where to place sensors, physical test programs could be reduced and digital twins better calibrated for operation (Kobryn 2023).

When building a representation of a fleet asset in a digital twin for maintenance and life-cycle predictions, it is important to capture the sources of manufacturing, operational, and environmental variation to understand how a particular component is operating in the field. This understanding enables the digital twin to have an appropriate fidelity to be useful in accurately predicting asset maintenance needs (Deshmukh 2023).

For DoD to move from digital twin "models to action," it is important to consider the following enablers: uncertainty propagation, fast inference, model error quantification, identifiability, causality, optimization and control, surrogates and reduced-order models, and multifidelity information. Integrating data science and domain knowledge is critical to enable decision-making based on analytics to drive process change. Managing massive amounts of data and applying advanced analytics with a new level of intelligent decision-making will be needed to fully take advantage of digital twins in the future. There is also a need for further research in ontologies and harmonization among the digital twin user community; interoperability (from cells, to units, to systems, to systems-of-systems); causality, correlation, and uncertainty quantification; data–physics fusion; and strategies to change the testing and organizational culture (Deshmukh 2023; Duraisamy 2023; Grieves 2023).

Opportunities exist in the national security arena to test, design, and prototype processes and exercise virtual prototypes in military campaigns or with geopolitical analysis to improve mission readiness (Bochenek 2023).

Digital Twin Examples, Needs, and Opportunities for Atmospheric and Climate Sciences

Digital twins are being explored and implemented in a variety of contexts within the atmospheric, climate, and sustainability sciences. Specific use cases and opportunities are presented in *Opportunities and Challenges for Digital Twins in Atmospheric and Climate Sciences: Proceedings of a Workshop—in Brief* in Appendix C (NASEM 2023a). Key messages from the workshop panelists and speakers are summarized here. Destination Earth, or DestinE, is a collaborative European effort to model the planet and capture both natural and human activities. Plans for DestinE include interactive simulations of Earth systems, improved prediction capabilities, support for policy decisions, and mechanisms for members of the broader community to engage with its data (European Commission 2023). The models enabling DestinE are intended to be more realistic and of higher resolution, and the digital twin will incorporate both real and synthetic data (Modigliani 2023). The infrastructure required to support such robust and large-scale atmospheric, climate, and sustainability digital twins, however, necessitates increased observational abilities, computational capacity, mechanisms for large-scale data handling, and federated resource management. Such large-scale digital twins necessitate increased computational capacity, given that significant capacity is required to resolve multiple models of varying scale. Moreover, increasing computational abilities is not sufficient; computational capacity must also be used efficiently.

It is important to note that climate predictions do not necessarily require real-time updates, but some climate-related issues, such as wildfire response planning, might (Ghattas 2023). Three specific thrusts could help to advance the sort of climate modeling needed to realize digital twins: research on parametric sparsity and generalizing observational data, generation of training data and computation for highest possible resolution, and uncertainty quantification and calibration based on both observational and synthetic data (Schneider 2023). ML could be used to expedite the data assimilation process of such diverse data.

There are many sources of unpredictability that limit the applicability of digital twins to atmospheric prediction or climate change projection. The atmosphere, for example, exhibits nonlinear behavior on many time scales. As a chaotic fluid that is sensitively dependent on initial conditions, the predictability of the atmosphere at instantaneous states is inherently limited. Similarly, the physics of the water cycle introduce another source of unpredictability. The water phase changes are associated with exchanges of energy, and they introduce irreversible conditions as water changes phase from vapor to liquid or solid in the atmosphere and precipitates out to the Earth's surface or the oceans.

The importance of—and challenges around—incorporating uncertainty into digital twins cannot be overstated. Approaches that rely on a Bayesian framework could help, as could utilizing reduced-order and surrogate models for tractability (Ghattas 2023) or utilizing fast sampling to better incorporate uncertainty (Balaji

2023). Giving users increased access to a digital twin's supporting data may foster understanding of the digital twin's uncertainty (McGovern 2023).

Establishing and maintaining confidence in and reliability of digital twins is critical for their use. One area for further development is tools that will assess the quality of a digital twin's outputs, thus bolstering confidence in the system (NASEM 2023a). Predicting extreme events also poses challenges for widespread digital twin development and adoption. Because extreme events are, by definition, in the tail end of a distribution, methods for validating extreme events and long-term climate predictions are needed.

It is important to note that digital twins are often designed to meet the needs of many stakeholders, often beyond the scientific community. Using physics-based models in conjunction with data-driven models can help to incorporate social justice factors into community-centric metrics (Di Lorenzo 2023). It is necessary to include diverse thinking in a digital twin and to consider the obstacles current funding mechanisms pose toward the cross-disciplinary work that would foster such inclusion (Asch 2023).

Digital Twin Examples, Needs, and Opportunities for Biomedical Applications

Many researchers hold that digital twins are not yet in practical use for decision-making in the biomedical space, but extensive work to advance their development is ongoing. Many of these efforts are described in *Opportunities and Challenges for Digital Twins in Biomedical Research: Proceedings of a Workshop—in Brief* in Appendix D (NASEM 2023b). The European Union has funded various projects for digital twins in the biomedical space. The European Virtual Human Twin (EDITH)[4] has the mission of creating a roadmap toward fully integrated multiscale and multiorgan whole-body digital twins. The goal of the project is to develop a cloud-based repository of digital twins for health care including data, models, algorithms, and good practices, providing a virtual collaboration environment. The team is also designing a simulation platform to support the transition toward an integrated twin. To prototype the platform, they have selected use cases in applications including cancer, cardiovascular disease, and osteoporosis. While questions for EDITH remain, including in the areas of technology (e.g., data, models, resource integration, infrastructure); users (e.g., access and workflows); ethics and regulations (e.g., privacy and policy); and sustainability (e.g., clinical uptake and business modeling) (Al-Lazikani et al. 2023), the work in this space is notable. DIGIPREDICT[5] and the Swedish Digital

---

[4] The website for the European Virtual Human Twin is https://www.edith-csa.eu, accessed June 30, 2023.

[5] The website for DIGIPREDICT is https://www.digipredict.eu, accessed June 30, 2023.

Twin Consortium[6] are two other examples of emerging European Union–funded projects working toward biomedical digital twins.

Technical challenges in modeling, computation, and data all pose current barriers to implementing digital twins for biomedical use. Because medical data are often sparse and collecting data can be invasive to patients, researchers need strategies to create working models despite missing data. A combination of data-driven and mechanistic models can be useful to this end (Glazier 2023; Kalpathy-Cramer 2023), but these approaches can remain limited due to the complexities and lack of understanding of the full biological processes even when sufficient data are available. In addition, data heterogeneity and the difficulty of integrating disparate multimodal data, collected across different time and size scales, also engender significant research questions. New techniques are necessary to harmonize, aggregate, and assimilate heterogenous data for biomedical digital twins (Koumoutsakos 2023; Sachs 2023). Furthermore, achieving interoperability and composability of models will be essential (Glazier 2023).

Accounting for uncertainty in biomedical digital twins as well as communicating and making appropriate decisions based on uncertainty will be vital to their practical application. As discussed more in Chapter 6, trust is paramount in the use of digital twins—and this is particularly critical for the use of these models in health care. Widespread adoption of digital twins will likely not be possible until patients, biologists, and clinicians trust them, which will first require education and transparency within the biomedical community (Enderling 2023; Miller 2023). Clear mechanisms for communicating uncertainty to digital twin users are a necessity. Though many challenges remain, opportunity also arises in that predictions from digital twins can open a line of communication between clinician and patient (Enderling 2023).

Ethical concerns are also important to consider throughout the process of developing digital twins for biomedical applications; these concerns cannot merely be an afterthought (NASEM 2023b). Bias inherent in data, models, and clinical processes needs to be evaluated and considered throughout the life cycle of a digital twin. Particularly considering the sensitive nature of medical data, it is important to prioritize privacy and security issues. Data-sharing mechanisms will also need to be developed, especially considering that some kinds of aggregate health data will never be entirely de-identifiable (Price 2023).

ADVANCING DIGITAL TWIN STATE OF THE ART REQUIRES AN INTEGRATED RESEARCH AGENDA

Despite the existence of examples of digital twins providing practical impact and value, the sentiment expressed across multiple committee information-gath-

---

[6] The website for the Swedish Digital Twin Consortium is https://www.sdtc.se, accessed June 30, 2023.

ering sessions is that the publicity around digital twins and digital twin solutions currently outweighs the evidence base of success. For example, in a briefing to the committee, Mark Girolami, chief scientist of The Alan Turing Institute, stated that the "Digital Twin evidence base of success and added value is seriously lacking" (Girolami 2022).

Conclusion 2-5: Digital twins have been the subject of widespread interest and enthusiasm; it is challenging to separate what is true from what is merely aspirational, due to a lack of agreement across domains and sectors as well as misinformation. It is important to separate the aspirational from the actual to strengthen the credibility of the research in digital twins and to recognize that serious research questions remain in order to achieve the aspirational.

Conclusion 2-6: Realizing the potential of digital twins requires an integrated research agenda that advances each one of the key digital twin elements and, importantly, a holistic perspective of their interdependencies and interactions. This integrated research agenda includes foundational needs that span multiple domains as well as domain-specific needs.

Recommendation 1: Federal agencies should launch new crosscutting programs, such as those listed below, to advance mathematical, statistical, and computational foundations for digital twins. As these new digital twin–focused efforts are created and launched, federal agencies should identify opportunities for cross-agency interactions and facilitate cross-community collaborations where fruitful. An interagency working group may be helpful to ensure coordination.

- *National Science Foundation (NSF).* NSF should launch a new program focused on mathematical, statistical, and computational foundations for digital twins that cuts across multiple application domains of science and engineering.
    - The scale and scope of this program should be in line with other multidisciplinary NSF programs (e.g., the NSF Artificial Intelligence Institutes) to highlight the technical challenge being solved as well as the emphasis on theoretical foundations being grounded in practical use cases.
    - Ambitious new programs launched by NSF for digital twins should ensure that sufficient resources are allocated to the solicitation so that the technical advancements are evaluated using real-world use cases and testbeds.
    - NSF should encourage collaborations across industry and academia and develop mechanisms to ensure that small and medium-sized industrial and academic institutions can also compete and be successful leading such initiatives.

- Ideally, this program should be administered and funded by multiple directorates at NSF, ensuring that from inception to sunset, real-world applications in multiple domains guide the theoretical components of the program.
- *Department of Energy (DOE).* DOE should draw on its unique computational facilities and large instruments coupled with the breadth of its mission as it considers new crosscutting programs in support of digital twin research and development. It is well positioned and experienced in large, interdisciplinary, multi-institutional mathematical, statistical, and computational programs. Moreover, it has demonstrated the ability to advance common foundational capabilities while also maintaining a focus on specific use-driven requirements (e.g., predictive high-fidelity models for high-consequence decision support). This collective ability should be reflected in a digital twin grand challenge research and development vision for DOE that goes beyond the current investments in large-scale simulation to advance and integrate the other digital twin elements, including the physical/virtual bidirectional interaction and high-consequence decision support. This vision, in turn, should guide DOE's approach in establishing new crosscutting programs in mathematical, statistical, and computational foundations for digital twins.
- *National Institutes of Health (NIH).* NIH should invest in filling the gaps in digital twin technology in areas that are particularly critical to biomedical sciences and medical systems. These include bioethics, handling of measurement errors and temporal variations in clinical measurements, capture of adequate metadata to enable effective data harmonization, complexities of clinical decision-making with digital twin interactions, safety of closed-loop systems, privacy, and many others. This could be done via new cross-institute programs and expansion of current programs such as the Interagency Modeling and Analysis Group.
- *Department of Defense (DoD).* DoD's Office of the Under Secretary of Defense for Research and Engineering should advance the application of digital twins as an integral part of the digital engineering performed to support system design, performance analysis, developmental and operational testing, operator and force training, and operational maintenance prediction. DoD should also consider using mechanisms such as the Multidisciplinary University Research Initiative and Defense Acquisition University to support research efforts to develop and mature the tools and techniques for the ap-

plication of digital twins as part of system digital engineering and model-based system engineering processes.
- *Other federal agencies.* Many federal agencies and organizations beyond those listed above can play important roles in the advancement of digital twin research. For example, the National Oceanic and Atmospheric Administration, the National Institute of Standards and Technology, and the National Aeronautics and Space Administration should be included in the discussion of digital twin research and development, drawing on their unique missions and extensive capabilities in the areas of data assimilation and real-time decision support.

As described earlier in this chapter, VVUQ is a key element of digital twins that necessitates collaborative and interdisciplinary investment.

Recommendation 2: Federal agencies should ensure that verification, validation, and uncertainty quantification (VVUQ) is an integral part of new digital twin programs. In crafting programs to advance the digital twin VVUQ research agenda, federal agencies should pay attention to the importance of (1) overarching complex multiscale, multiphysics problems as catalysts to promote interdisciplinary cooperation; (2) the availability and effective use of data and computational resources; (3) collaborations between academia and mission-driven government laboratories and agencies; and (4) opportunities to include digital twin VVUQ in educational programs. Federal agencies should consider the Department of Energy Predictive Science Academic Alliance Program as a possible model to emulate.

KEY GAPS, NEEDS, AND OPPORTUNITIES

In Table 2-1, the committee highlights key gaps, needs, and opportunities across the digital twin landscape. This is not meant to be an exhaustive list of all opportunities presented in the chapter. For the purposes of this report, prioritization of a gap is indicated by 1 or 2. While the committee believes all of the gaps listed are of high priority, gaps marked 1 may benefit from initial investment before moving on to gaps marked with a priority of 2.

TABLE 2-1 Key Gaps, Needs, and Opportunities Across the Digital Twin Landscape

Maturity	Priority
Early and Preliminary Stages	
Development and deployment of digital twins that enable decision-makers to anticipate and adapt to evolving threats, plan and execute emergency response, and assess impact are needed.	2
Building trust is a critical step toward clinical integration of digital twins and in order to start building trust, methods to transparently and effectively communicate uncertainty quantification to all stakeholders are critical.	1
Privacy and ethical considerations must be made through the development, implementation, and life cycle of biomedical digital twins, including considerations of biases in the data, models, and accepted clinical constructs and dogmas that currently exist.	2
Some Research Base Exists But Additional Investment Required	
Additional work is needed to advance scalable algorithms in order to bring digital twins to fruition at the Department of Defense. Specific examples of areas of need include uncertainty quantification, fast inference, model error quantification, identifiability, causality, optimization and control, surrogates and reduced-order models, multifidelity approaches, ontologies, and interoperability. The scalability of machine learning algorithms in uncertainty quantification settings is a significant issue. The computational cost of applying machine learning to large, complex systems, especially in an uncertainty quantification context, needs to be addressed.	1
Digital twins for defense applications require mechanisms and infrastructure to handle large quantities of data. This is a need that is common to digital twins across many domains, but the nature of data for defense applications brings some unique challenges due to the need for classified handling of certain sensor data and the need for near-real-time processing of the data to allow for minimal reaction time.	1
Large-scale atmospheric, climate, and sustainability digital twins must be supported by increased observational abilities, more efficient use of computational capacity, effective data handling, federated resource management, and international collaboration.	1
Methods for validating atmospheric, climate, and sustainability sciences digital twin predictions over long horizons and extreme events are needed.	1
Mechanisms to better facilitate cross-disciplinary collaborations are needed to achieve inclusive digital twins for atmospheric, climate, and sustainability sciences.	2
Due to the heterogeneity, complexity, multimodality, and breadth of biomedical data, the harmonization, aggregation, and assimilation of data and models to effectively combine these data into biomedical digital twins require significant technical research.	1
Research Base Exists with Opportunities to Advance Digital Twins	
Uncertainty quantification is critical to digital twins for atmospheric, climate, and sustainability sciences and will generally require surrogate models and/or improved sampling techniques.	2

REFERENCES

AIAA (American Institute of Aeronautics and Astronautics) Computational Fluid Dynamics Committee. 1998. *Guide for the Verification and Validation of Computational Fluid Dynamics Simulations.* Reston, VA.

AIAA Digital Engineering Integration Committee. 2020. "Digital Twin: Definition and Value." AIAA and AIA Position Paper.

Al-Lazikani, B., G. An, and L. Geris. 2023. "Connecting Across Scales." Presentation to the Workshop on Opportunities and Challenges for Digital Twins in Biomedical Sciences. January 30. Washington, DC.

Asch, M. 2023. Presentation to the Workshop on Digital Twins in Atmospheric, Climate, and Sustainability Science. February 2. Washington, DC.

Balaji, V. 2023. "Towards Traceable Model Hierarchies." Presentation to the Workshop on Digital Twins in Atmospheric, Climate, and Sustainability Science. February 1. Washington, DC.

Bennett, H., M. Birkin, J. Ding, A. Duncan, and Z. Engin. 2023. "Towards Ecosystems of Connected Digital Twins to Address Global Challenges." White paper. London, England: The Alan Turing Institute.

Bochenek, G. 2023. Presentation to the Workshop on Opportunities and Challenges for Digital Twins in Engineering. February 9. Washington, DC.

Center for A.I. Safety. 2023. "Statement on AI Risk [open letter]."

Darema, F. 2004. "Dynamic Data-Driven Applications Systems: A New Paradigm for Application Simulations and Measurements." *Lecture Notes in Computer Science* 3038:662–669.

Deshmukh, D. 2023. Presentation to the Workshop on Opportunities and Challenges for Digital Twins in Engineering. February 7. Washington, DC.

Di Lorenzo, E. 2023. Presentation to the Workshop on Digital Twins in Atmospheric, Climate, and Sustainability Science. February 2. Washington, DC.

Duraisamy, K. 2023. Presentation to the Workshop on Opportunities and Challenges for Digital Twins in Engineering. February 7. Washington, DC.

Enderling, H. 2023. Presentation to the Workshop on Opportunities and Challenges for Digital Twins in Biomedical Sciences. January 30. Washington, DC.

European Commission. 2023. "Destination Earth." https://digital-strategy.ec.europa.eu/en/policies/destination-earth. Last modified April 20.

Gahn, M.S. 2023. Presentation to the Workshop on Opportunities and Challenges for Digital Twins in Engineering. February 7. Washington, DC.

Ghattas, O. 2023. Presentation to the Workshop on Digital Twins in Atmospheric, Climate, and Sustainability Science. February 1. Washington, DC.

Girolami, M. 2022. "Digital Twins: Essential, Mathematical, Statistical and Computing Research Foundations." Presentation to the Committee on Foundational Research Gaps and Future Directions for Digital Twins. November 21. Washington, DC.

Glaessgen, E., and D. Stargel. 2012. "The Digital Twin Paradigm for Future NASA and US Air Force Vehicles." AIAA Paper 2012-1818 in *Proceedings of the 53rd AIAA/ASME/ASCE/AHS/ASC Structures, Structural Dynamics and Materials Conference.* April. Honolulu, Hawaii.

Glazier, J.A. 2023. Presentation to the Workshop on Opportunities and Challenges for Digital Twins in Biomedical Sciences. January 30. Washington, DC.

Goodchild, M. 2023. Presentation to the Workshop on Digital Twins in Atmospheric, Climate, and Sustainability Science. February 2. Washington, DC.

Grieves, M. 2005a. *Product Lifecycle Management: Driving the Next Generation of Lean Thinking.* New York: McGraw-Hill.

Grieves, M. 2005b. "Product Lifecycle Management: The New Paradigm for Enterprises." *International Journal of Product Development* 2 1(2):71–84.

Grieves, M. 2023. Presentation to the Workshop on Opportunities and Challenges for Digital Twins in Engineering. February 7. Washington, DC.

Hendrickson, B., B. Bland, J. Chen, P. Colella, E. Dart, J. Dongarra, T. Dunning, et al. 2020. *ASCR@ 40: Highlights and Impacts of ASCR's Programs*. Department of Energy Office of Science.

Kalpathy-Cramer, J. 2023. "Digital Twins at the Organ, Tumor, and Microenvironment Scale." Presentation to the Workshop on Opportunities and Challenges for Digital Twins in Biomedical Sciences. January 30. Washington, DC.

Kobryn, P. 2023. "AFRL Airframe Digital Twin." Presentation to the Workshop on Opportunities and Challenges for Digital Twins in Engineering. February 9. Washington, DC.

Koumoutsakos, P. 2023. Presentation to the Workshop on Opportunities and Challenges for Digital Twins in Biomedical Sciences. January 30. Washington, DC.

Mangravite, L. 2023. Presentation to the Workshop on Opportunities and Challenges for Digital Twins in Biomedical Sciences. January 30. Washington, DC.

McGovern, A. 2023. Presentation to the Workshop on Digital Twins in Atmospheric, Climate, and Sustainability Science. February 2. Washington, DC.

Miller, D. 2023. "Prognostic Digital Twins in Practice." Presentation to the Workshop on Opportunities and Challenges for Digital Twins in Biomedical Sciences. January 30. Washington, DC.

Modigliani, U. 2023. "Earth System Digital Twins and the European Destination Earth Initiative." Presentation to the Workshop on Digital Twins in Atmospheric, Climate, and Sustainability Science. February 1. Washington, DC.

NASEM (National Academies of Sciences, Engineering, and Medicine). 2023a. *Opportunities and Challenges for Digital Twins in Atmospheric and Climate Sciences: Proceedings of a Workshop—in Brief*. Washington, DC: The National Academies Press.

NASEM. 2023b. *Opportunities and Challenges for Digital Twins in Biomedical Research: Proceedings of a Workshop—in Brief*. Washington, DC: The National Academies Press.

NASEM. 2023c. *Opportunities and Challenges for Digital Twins in Engineering: Proceedings of a Workshop—in Brief*. Washington, DC: The National Academies Press.

Niederer, S.A., M.S. Sacks, M. Girolami, and K. Willcox. 2021. "Scaling Digital Twins from the Artisanal to the Industrial." *Nature Computational Science* 1(5):313–320.

NIST (National Institute of Standards and Technology). 2009. "The System Development Life Cycle (SDLC)." *ITL Bulletin*. https://tsapps.nist.gov/publication/get_pdf.cfm?pub_id=902622.

NRC (National Research Council). 2012. *Assessing the Reliability of Complex Models: Mathematical and Statistical Foundations of Verification, Validation, and Uncertainty Quantification*. Washington, DC: The National Academies Press.

Piascik, B., J. Vickers, D. Lowry, S. Scotti, J. Stewart, and A. Calomino. 2012. "Materials, Structures, Mechanical Systems, and Manufacturing Roadmap: Technology Area 12." Washington, DC: NASA.

Price, N. 2023. Presentation to the Workshop on Opportunities and Challenges for Digital Twins in Biomedical Sciences. January 30. Washington, DC.

Rawlings, J.B., D.Q. Mayne, and M. Diehl. 2017. *Model Predictive Control: Theory, Computation, and Design*. Vol. 2. Madison, WI: Nob Hill Publishing.

Reichle, R.H. 2008. "Data Assimilation Methods in the Earth Sciences." *Advances in Water Resources* 31(11):1411–1418.

Roose, K. 2023. "A.I. Poses 'Risk of Extinction,' Industry Leaders Warn." *New York Times*. May 30. https://www.nytimes.com/2023/05/30/technology/ai-threat-warning.html.

Sachs, J.R. 2023. "Digital Twins: Pairing Science with Simulation for Life Sciences." Presentation to the Workshop on Opportunities and Challenges for Digital Twins in Biomedical Sciences. January 30. Washington, DC.

Schneider, T. 2023. Presentation to the Workshop on Digital Twins in Atmospheric, Climate, and Sustainability Science. February 1. Washington, DC.

Shafto, M., M. Conroy, R. Doyle, E. Glaessgen, C. Kemp, J. LeMoigne, and L. Wang. 2012. "Modeling, Simulation, Information Technology and Processing Roadmap." *National Aeronautics and Space Administration* 32(2012):1–38.

Shanley, L. 2023. "Discussion of Ethical Considerations of Digital Twins." Presentation to the Committee on Foundational Research Gaps and Future Directions for Digital Twins. April 27. Washington, DC.

Tuegel, E.J., A.R. Ingraffea, T.G. Eason, and S.M. Spottswood. 2011. "Reengineering Aircraft Structural Life Prediction Using a Digital Twin." *International Journal of Aerospace Engineering* 2011:1–14.

Wells, J. 2022. "Digital Twins and NVIDIA Omniverse." Presentation to the Committee on Foundational Research Gaps and Future Directions for Digital Twins. November 21. Washington, DC.

3

Virtual Representation: Foundational Research Needs and Opportunities

The digital twin virtual representation comprises a computational model or set of coupled models. This chapter identifies research needs and opportunities associated with creating, scaling, validating, and deploying models in the context of a digital twin. The chapter emphasizes the importance of the virtual representation being fit for purpose and the associated needs for data-centric and model-centric formulations. The chapter discusses multiscale modeling needs and opportunities, including the importance of hybrid modeling that combines mechanistic models and machine learning (ML). This chapter also discusses the challenges of integrating component and subsystem digital twins into the virtual representation. Surrogate modeling needs and opportunities for digital twins are also discussed, including surrogate modeling for high-dimensional, complex multidisciplinary systems and the essential data assimilation, dynamic updating, and adaptation of surrogate models.

FIT-FOR-PURPOSE VIRTUAL REPRESENTATIONS FOR DIGITAL TWINS

As discussed in Chapter 2, the computational models underlying the digital twin virtual representation can take many mathematical forms (including dynamical systems, differential equations, and statistical models) and need to be "fit for purpose" (meaning that model types, fidelity, resolution, parameterization, and quantities of interest must be chosen and potentially dynamically adapted to fit the particular decision task and computational constraints). The success of a digital twin hinges critically on the availability of models that can represent the physical counterpart with fidelity that is fit for purpose, and that can be used to issue predictions with known confidence, possibly in extrapolatory regimes, all while satisfying computational resource constraints.

As the foundational research needs and opportunities for modeling in support of digital twins are outlined, it is important to emphasize that there is no one-size-fits-all approach. The vast range of domain applications and use cases that are envisioned for digital twins requires a similarly vast range of models: first-principles, mechanistic, and empirical models all have a role to play.

There are several areas in which the state of the art in modeling is currently a barrier to achieving the impact of digital twins, due to the challenges of modeling complex multiphysics systems across multiple scales. In some cases, the mathematical models are well understood, and these barriers relate to our inability to bridge scales in a computationally tractable way. In other cases, the mathematical models are lacking, and discovery of new models that explain observed phenomena is needed. In yet other cases, mathematical models may be well understood and computationally tractable to solve at the component level, but foundational questions remain around stability and accuracy when multiple models are coupled at a full system or system-of-systems level. There are other areas in which the state of the art in modeling provides potential enablers for digital twins. The fields of statistics, ML, and surrogate modeling have advanced considerably in recent years, but a gap remains between the class of problems that has been addressed and the modeling needs for digital twins.

Some communities focus on high-fidelity models in the development of digital twins while others define digital twins using simplified and/or surrogate models. Some literature states that a digital twin must be a high-resolution, high-fidelity replica of the physical system (Bauer et al. 2021; NASEM 2023a). An early definition of a digital twin proposed "a set of virtual information constructs that fully describes a potential or actual physical manufactured product from the micro atomic level to the macro geometrical level. At its optimum, any information that could be obtained from inspecting a physical manufactured product can be obtained from its Digital Twin" (Grieves 2014). Other literature proposes surrogate modeling as a key enabler for digital twins (Hartmann et al. 2018; NASEM 2023c), particularly recognizing the dynamic (possibly real-time) nature of many digital twin calculations.

Conclusion 3-1: A digital twin should be defined at a level of fidelity and resolution that makes it fit for purpose. Important considerations are the required level of fidelity for prediction of the quantities of interest, the available computational resources, and the acceptable cost. This may lead to the digital twin including high-fidelity, simplified, or surrogate models, as well as a mixture thereof. Furthermore, a digital twin may include the ability to represent and query the virtual models at variable levels of resolution and fidelity depending on the particular task at hand and the available resources (e.g., time, computing, bandwidth, data).

Determining whether a virtual representation is fit for purpose is itself a mathematical gap when it comes to the complexity of situations that arise with digital twins. For a model to be fit for purpose, it must balance the fidelity of predictions of quantities of interest with computational constraints, factoring in acceptable levels of uncertainty to drive decisions. If there is a human in the digital twin loop, fitness for purpose must also account for human–digital twin interaction needs such as visualization and communication of uncertainty. Furthermore, since a digital twin's purpose may change over time, the requirements for it to be fit for purpose may also evolve. Historically, computational mathematics has addressed accuracy requirements for numerical solution of partial differential equations using rigorous approaches such as a posteriori error estimation combined with numerical adaptivity (Ainsworth and Oden 1997). These kinds of analyses are an important ingredient of assessing fitness for purpose; however, the needs for digital twins go far beyond this, particularly given the range of model types that digital twins will employ and the likelihood that a digital twin will couple multiple models of differing fidelity. A key feature for determining fitness for purpose is assessing whether the fusion of a mathematical model, potentially corrected via a discrepancy function, and observational data provides relevant information for decision-making. Another key aspect of determining digital twin fitness for purpose is assessment of the integrity of the physical system's observational data, as discussed in Chapter 4.

Finding 3-1: Approaches to assess modeling fidelity are mathematically mature for some classes of models, such as partial differential equations that represent one discipline or one component of a complex system; however, theory and methods are less mature for assessing the fidelity of other classes of models (particularly empirical models) and coupled multiphysics, multi-component systems.

An additional consideration in determining model fitness for purpose is the complementary role of models and data—a digital twin is distinguished from traditional modeling and simulation in the way that models and data work together to drive decision-making. Thus, it is important to analyze the entire digital twin ecosystem when assessing modeling needs and the trade-offs between data-driven and model-driven approaches (Ferrari 2023).

In some cases, there is an abundance of data, and the decisions to be made fall largely within the realm of conditions represented by the data. In these cases, a data-centric view of a digital twin (Figure 3-1) is appropriate—the data form the core of the digital twin, the numerical model is likely heavily empirical (e.g., obtained via statistical or ML methods), and analytics and decision-making wrap around this numerical model. An example of such a setting is the digital twin of an aircraft engine, trained on a large database of sensor data and flight logs collected across a fleet of engines (*Aviation Week Network* 2019; Sieger 2019). Other cases are data-poor, and the digital twin will be called on to issue predictions in extrapolatory regimes that go well beyond the available data. In these cases, a model-centric view of a digital twin (Figure 3-1) is appropriate—a mathematical model and its associated numerical model form the core of the digital twin, and data are assimilated through the lens of these models. Examples include climate digital twins, where observations are typically spatially sparse and predictions may extend decades into the future (NASEM 2023a), and cancer patient digital twins, where observations are typically temporally sparse and the increasingly patient-specific and complex nature of diseases and therapies requires predictions of patient responses that go beyond available data (Yankeelov 2023). In these data-poor situations, the models play a greater role in determining digital twin fidelity. As discussed in the next section, an important need is to advance hybrid modeling approaches that leverage the synergistic strengths of data-driven and model-driven digital twin formulations.

MULTISCALE MODELING NEEDS AND OPPORTUNITIES FOR DIGITAL TWINS

A fundamental challenge for digital twins is the vast range of spatial and temporal scales that the virtual representation may need to address. The following section describes research opportunities for modeling across scales in support of digital twins and the need to integrate empirical and mechanistic methods for

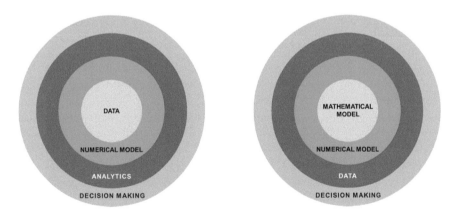

FIGURE 3-1 Conceptualizing a digital twin: data-centric and model-centric views. In data-rich settings, the data form the core of the digital twin, while in data-poor settings, mathematical models play a more important role.
SOURCE: Courtesy of Karen Willcox.

hybrid approaches to leverage the best of both data-driven and model-driven digital twin formulations.

The Predictive Power of Digital Twins Requires Modeling Across Scales

For many applications, the models that underlie the digital twin virtual representation must represent the behavior of the system across a wide range of spatial and temporal scales. For systems with a wide range of scales on which there are significant nonlinear scale interactions, it may be impossible to represent explicitly in a digital model the full richness of behavior at all scales and including all interactions. For example, the Earth's atmosphere and oceans are components of the Earth system, and their instantaneous and statistical behaviors are described respectively as weather and climate. These behaviors exhibit a wide range of variability on both spatial scales (from millimeters to tens of thousands of kilometers) and temporal scales (from seconds to centuries). Similarly, relevant dynamics in biological systems range from nanometers to meters in spatial scales and from milliseconds to years in temporal scales. In biomedical systems, modeling requirements range across scales from the molecular to the whole-body physiology and pathophysiology to populations. Temporal ranges in nanoseconds represent biochemical reactions, signaling pathways, gene expression, and cellular processes such as redox reactions or transient protein modifications. These events underpin the larger-scale interactions between cells, tissues, and organs; multiple organs and systems converge to address disease and non-disease states.

Numerical models of many engineering systems in energy, transportation, and aerospace sectors also span a range of temporal and spatial resolutions, and complexity owing to multiphysics phenomena (e.g., chemical reactions, heat transfer, phase change, unsteady flow/structure interactions) and resolution of intricate geometrical features. In weather and climate simulations, as well as in many engineered and biomedical systems, system behavior is explicitly modeled across a limited range of scales—typically, from the largest scale to an arbitrary cutoff scale determined by available modeling resources—and the remaining (small) scales are represented in a parameterized form. Fortunately, in many applications, the smaller unresolved scales are known to be more universal than the large-scale features and thus more amenable to phenomenological parameterization. Even so, a gap remains between the scales that can be simulated and actionable scales.

An additional challenge is that as finer scales are resolved and a given model achieves greater fidelity to the physical counterpart it simulates, the computational and data storage/analysis requirements increase. This limits the applicability of the model for some purposes, such as uncertainty quantification, probabilistic prediction, scenario testing, and visualization. As a result, the demarcation between resolved and unresolved scales is often determined by computational constraints rather than a priori scientific considerations. Another challenge to increasing resolution is that the scale interactions may enter a different regime as scales change. For example, in atmospheric models, turbulence is largely two-dimensional at scales larger than 10 km and largely three-dimensional at scales smaller than 10 km; the behavior of fluid-scale interactions fundamentally changes as the model grid is refined.

Thus, there are incentives to drive modeling for digital twins in two directions: toward resolution of finer scales to achieve greater realism and fidelity on the one hand, and toward simplifications to achieve computational tractability on the other. There is a motivation to do both by increasing model resolution to acquire data from the most realistic possible model that can then be mined to extract a more tractable model that can be used as appropriate.

Finding 3-2: Different applications of digital twins drive different requirements for modeling fidelity, data, precision, accuracy, visualization, and time-to-solution, yet many of the potential uses of digital twins are currently intractable to realize with existing computational resources.

Finding 3-3: Often, there is a gap between the scales that can be simulated and actionable scales. It is necessary to identify the intersection of simulated and actionable scales in order to support optimizing decisions. The demarcation between resolved and unresolved scales is often determined by available computing resources, not by a priori scientific considerations.

Recommendation 3: In crafting research programs to advance the foundations and applications of digital twins, federal agencies should create mechanisms to provide digital twin researchers with computational resources, recognizing the large existing gap between simulated and actionable scales and the differing levels of maturity of high-performance computing across communities.

Finding 3-4: Advancing mathematical theory and algorithms in both data-driven and multiscale physics-based modeling to reduce computational needs for digital twins is an important complement to increased computing resources.

Hybrid Modeling Combining Mechanistic Models and Machine Learning

Hybrid modeling approaches—synergistic combinations of empirical and mechanistic modeling approaches that leverage the best of both data-driven and model-driven formulations—were repeatedly emphasized during this study's information gathering (NASEM 2023a,b,c). This section provides some examples of how hybrid modeling approaches can address digital twin modeling challenges.

In biology, modeling organic living matter requires the integration of biological, chemical, and even electrical influences that stimulate or inhibit the living material response. For many biomedical applications, this requires the incorporation of smaller-scale biological phenomena that influence the dynamics of the larger-scale system and results in the need for multiphysics, multiscale modeling. Incorporating multiple smaller-scale phenomena allows modelers to observe the impact of these underlying mechanisms at a larger scale, but resolving the substantial number of unknown parameters to support such an approach is challenging. Data-driven modeling presents the ability to utilize the growing volume of biological and biomedical data to identify correlations and generate inferences about the behavior of these biological systems that can be tested experimentally. This synergistic use of data-driven and multiscale modeling approaches in biomedical and related fields is illustrated in Figure 3-2.

Advances in hybrid modeling in the Earth sciences are following similar lines. Models for weather prediction or climate simulation must solve multiscale and multiphysics problems that are computationally intractable at the necessary level of fidelity, as described above. Over the past several decades of work in developing atmospheric, oceanic, and Earth system models, the unresolved scales have been represented by parameterizations that are based on conceptual models of the relevant unresolved processes. With the explosion of Earth system observations from remote sensing platforms in recent years, this approach has been modified to incorporate ML methods to relate the behavior of unresolved processes to that of resolved processes. There are also experiments in replacing entire Earth system components with empirical artificial intelligence (AI) components. Furthermore, the use of ensemble modeling to approximate probability distributions invites the use of ML techniques, often in a Bayesian framework, to cull ensemble members that are less accurate or to define clusters of solutions that simplify the application to decision-making.

In climate and engineering applications, the potential for hybrid modeling to underpin digital twins is significant. In addition to modeling across scales as described above, hybrid models can help provide understandability and explainability. Often, a purely data-driven model can identify a problem or potential opportunity without offering an understanding of the root cause. Without this understanding, decisions related to the outcome may be less useful. The combination of data and mechanistic models comprising a hybrid model can help mitigate this problem. The aerospace industry has developed hybrid digital twin solutions that can analyze large, diverse data sets associated with part failures in aircraft engines using the data-driven capabilities of the hybrid model (Deshmukh 2022). Additionally, these digital twin solutions can provide root cause analysis indicators using the mechanistic-driven capabilities of the hybrid model.

However, there are several gaps in hybrid modeling approaches that need to be addressed to realize the full potential value of these digital twin solutions. These gaps exist in five major areas: (1) data quality, availability, and affordabil-

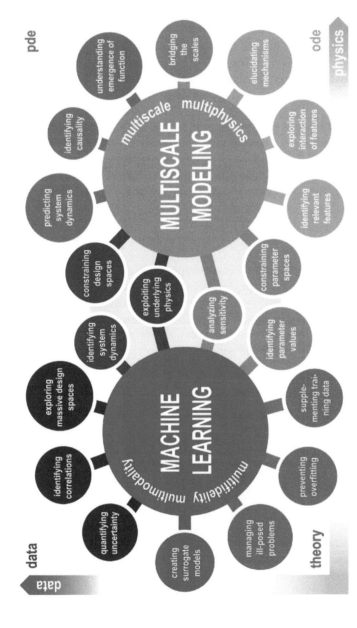

FIGURE 3-2 Machine learning and multiscale modeling approaches can be used separately and together to approach many aspects of the virtual representation of a digital twin.

NOTE: ODE, ordinary differential equation; PDE, partial differential equation.

SOURCE: M. Alber, A. Buganza Tepole, W.R. Cannon, et al., 2019, "Integrating Machine Learning and Multiscale Modeling—Perspectives, Challenges, and Opportunities in the Biological, Biomedical, and Behavioral Sciences," *npj Digital Medicine* 2(115), https://doi.org/10.1038/s41746-019-0193-y. Copyright 2019. CC BY 4.0.

ity; (2) model coupling and integration; (3) model validation and calibration; (4) uncertainty quantification and model interpretability; and (5) model scalability and management.

Data quality, availability, and affordability can be challenging in biomedical, climate, and engineering applications as obtaining accurate and representative data for model training and validation at an affordable price is difficult. Prior data collected may have been specific to certain tasks, limited by the cost of capture and storage, or deemed unsuitable for current use due to evolving environments and new knowledge. Addressing data gaps based on the fit-for-purpose requirements of the digital twin and an analysis of current available data is crucial. Minimizing the need for large sample sizes and designing methodologies to learn robustly from data sets with few samples would also help overcome these barriers. AI methods might be developed to predict a priori what amount and type of data are needed to support the virtual counterpart.

Combining data-driven models with mechanistic models requires effective coupling techniques to facilitate the flow of information (data, variables, etc.) between the models while understanding the inherent constraints and assumptions of each model. Coupling is complex in many cases, and model integration is even more so as it involves creating a single comprehensive model that represents the features and behaviors of both the data-driven and the mechanistic-driven model within a coherent framework. Both integration and coupling techniques require harmonizing different scales, assumptions, constraints, and equations, and understanding their implications on the uncertainty associated with the outcome. Matching well-known, model-driven digital twin representations with uncharacterized data-driven models requires attention to how the various levels of fidelity comprised in these models interact with each other in ways that may result in unanticipated overall digital twin behavior and inaccurate representation at the macro level. Another gap lies in the challenge of choosing the specific data collection points to adequately represent the effects of the less-characterized elements and augment the model-driven elements without oversampling the behavior already represented in the model-driven representations. Finally, one can have simulations that produce a large data set (e.g., a space-time field where each solution field is of high dimension) but only relatively few ensembles. In such cases, a more structured statistical model may be required to combine simulations and observations.

Model validation is another evident gap that needs to be overcome given the diverse nature of the involved data-driven and mechanistic models and their underlying assumptions. Validating data-driven models heavily relies on having sufficient and representative validation data for training as well as evaluating the accuracy of the outcome and the model's generalizability to new data. On the other hand, mechanistic-driven models heavily rely on calibration and parameter estimation to accurately reproduce against experimental and independent data. The validation and calibration processes for these hybrid models must be harmonized to ensure the accuracy and reliability required in these solutions.

Uncertainty quantification and model explainability and interpretability are significant gaps associated with hybrid systems. These systems must accurately account for uncertainties arising from both the data-driven and mechanistic-driven components of the model. Uncertainties can arise from various factors related to both components, including data limitations and quality, model assumptions, and parameter estimation. Addressing how these uncertainties are quantified and propagated through the hybrid model is another gap that must be tackled for robust predictions. Furthermore, interpreting and explaining the outcomes may pose a significant challenge, particularly in complex systems.

Finally, many hybrid models associated with biomedical, climate, and engineering problems can be computationally demanding and require unique skill sets. Striking a balance between techniques that manage the computational complexity of mechanistic models (e.g., parallelization and model simplification) and techniques used in data-driven models (e.g., graphics processing unit coding, pruning, and model compression) is essential. Furthermore, hybrid approaches require that domain scientists either learn details of computational complexity and data-driven techniques or partner with additional researchers to experiment with hybrid digital twins. Resolving how to achieve this combination and balance at a feasible and affordable level is a gap that needs to be addressed. Additionally, the model will need to be monitored and updated as time and conditions change and errors in the system arise, requiring the development of model management capabilities.

While hybrid modeling provides an attractive path forward to address digital twin modeling needs, simply crafting new hybrid models that better match available data is insufficient. The development of hybrid modeling approaches for digital twins requires rigorous verification, validation, and uncertainty quantification (VVUQ), including the quantification of uncertainty in extrapolatory conditions. If the hybrid modeling is done in a way that the data-driven components of the model are continually updated, then these updating methods also require associated VVUQ. Another challenge is that in many high-value contexts, digital twins need to represent both typical operating conditions and anomalous operating conditions, where the latter may entail rare or extreme events. As noted in Conclusion 2-2, a gap exists between the class of problems that has been considered in VVUQ for traditional modeling and simulation settings and the VVUQ problems that will arise for digital twins. Hybrid models—in particular those that infuse some form of black-box deep learning—represent a particular gap in this regard.

Finding 3-5: Hybrid modeling approaches that combine data-driven and mechanistic modeling approaches are a productive path forward for meeting the modeling needs of digital twins, but their effectiveness and practical use are limited by key gaps in theory and methods.

INTEGRATING COMPONENT AND SUBSYSTEM DIGITAL TWINS

The extent to which the virtual representation will integrate component and subsystem models is an important consideration in modeling digital twins. A digital twin of a system of systems will likely couple multiple constituent digital twins. Integration of models and data to this extent goes beyond what is done routinely and entails a number of foundational mathematical and computational challenges. In addition to the software challenge of coupling models and solvers, VVUQ tasks and the determination of fitness for purpose become much more challenging in the coupled setting.

Modeling of a complex system often requires coupling models of different components/subsystems of the system, which presents additional challenges beyond modeling of the individual components/subsystems. For example, Earth system models couple models of atmosphere, land surface, river, ocean, sea ice, and land ice to represent interactions among these subsystems that determine the internal variability of the system and its response to external forcing. Component models that are calibrated individually to be fit for purpose when provided with observed boundary conditions of the other components may behave differently when the component models are coupled together due to error propagation and nonlinear feedback between the subsystems. This is particularly the case when models representing the different components/subsystems have different fidelity or mathematical forms, necessitating the need for additional mathematical operations such as spatiotemporal filtering, which adds uncertainty in the coupled model.

Another example is the coupling of human system models with Earth system models, which often differ in model fidelity as well as in mathematical forms. Furthermore, in the context of digital twins, some technical challenges remain in coupled model data assimilation, such as properly initializing each component model. Additional examples of the integration of components are shown in Box 3-1.

Interoperability of software and data are a challenge across domains and pose a particular challenge when integrating component and subsystem digital twins. Semantic and syntactic interoperability, in which data are exchanged between and understood by the different systems, can be challenging given the possible difference in the systems. Furthermore, assumptions made in one model can be distinct from the assumptions made in other models. Some communities have established approaches to reducing interoperability—for example, though the use of shared standards for data, software, and models, or through the use of software templates—and this is a critical aspect of integrating complex digital twin models.

Finding 3-6: Integration of component/subsystem digital twins is a pacing item for the digital twin representation of a complex system, especially if different fidelity models are used in the digital twin representation of its components/subsystems.

BOX 3-1
Examples of the Integration of Components

Gas Turbine Engine

Gas turbines are used as propulsion devices in aviation and for electric power generation. Numerical simulation of the aerothermal flow through the entire gas turbine engine involves many different physical processes, which are described using different models and even different computer codes. Simulation of different modules (e.g., compressors, combustor, and turbines) separately requires imposition of (artificial) boundary conditions at the component interfaces, which are not known a priori in detail and can result in missing crucial interactions between components such as thermoacoustic instabilities. In an integrated simulation, the interaction between modules requires exchange of information between the participating solvers. Automation of this exchange requires a coupler software that manages the required exchange of information between the solvers in a seamless and efficient manner.[a]

Human Cardiac System

Integrated simulation of blood flow through the cardiac system involves a range of parameters. Models can capture genetic base characteristics, genetic variations, gene expression, and molecular interactions at the cellular and tissue levels to understand how specific genetic factors influence physiological processes and disease susceptibility. Structural information collected by imaging technology (e.g., magnetic resonance imaging, computed tomography scans) provides anatomical orientation of chambers, valves, and major blood vessels. Electrical activity of the heart captures the generation and propagation of electrical signals that coordinate the contraction of cardiac muscle cells. Models based on the Hodgkin-Huxley equations or other electrophysiological models are utilized to replicate the cardiac action potential and activation patterns. Mechanical aspects involve modeling the contraction and relaxation of cardiac muscle cells using parameters such as ventricular pressure, myocardial deformation, and valve dynamics. Hemodynamic models use computational fluid dynamics to simulate blood flow within the cardiac system (blood pressure, flow rates, and resistance), accounting for the interaction between the heart and the vasculature. Techniques can be employed to simulate blood flow patterns. Modeling the interaction between blood flow and the heart tissue captures the effects of fluid-structure interaction. The digital twin can incorporate regulatory mechanisms that control heart rate, blood pressure, and other physiological variables that maintain homeostasis and response mechanisms. However, each of these parameters is subject to multiple uncertainties: physiological or genetic parameters may vary between individuals; input data may be unreliable (e.g., imaging resolution); and experimental validation may contain measurement noise or other capture limitations.

---

[a] J. Alonso, S. Hahn, F. Ham, M. Herrmann, G. Iaccarino, G. Kalitzin, P. LeGresley, et al., 2006, "CHIMPS: A High-Performance Scalable Module for Multiphysics Simulations," *Aerospace Research Council* 1–28.

SURROGATE MODELING NEEDS AND OPPORTUNITIES FOR DIGITAL TWINS

Surrogate models play a key role in addressing the computational challenges of digital twins. Surrogate models can be categorized into three types: statistical data-fit models, reduced-order models, and simplified models.

- *Statistical data-fit models* use statistical methods to fit approximate input-output maps to training data, with the surrogate model employing a generic functional form that does not explicitly reflect the structure of the physical governing equations underlying the numerical simulations.
- *Reduced-order models* incorporate low-dimensional structure learned from training data into a structured form of the surrogate model that reflects the underlying physical governing equations.
- *Simplified models* are obtained in a variety of ways, such as coarser grids, simplified physical assumptions, and loosened residual tolerances.

Surrogate modeling is a broad topic, with many applications beyond digital twins. This section focuses on unique challenges that digital twins pose to surrogate modeling and the associated foundational gaps in surrogate modeling methods. A first challenge is the scale at which surrogate modeling will be needed. Digital twins by their nature may require modeling at the full system scale, with models involving multiple disciplines, covering multiple system components, and described by parameter spaces of high dimensions. A second challenge is the critical need for VVUQ of surrogate models, recognizing the uncertain conditions under which digital twins will be called on to make predictions, often in extrapolatory regimes. A third challenge relates to the dynamic updating and adaptation that is key to the digital twin concept. Each one of these challenges highlights gaps in the current state of the art in surrogate modeling, as the committee discusses in more detail in the following.

Surrogate modeling is an enabler for computationally efficient digital twins, but there is a limited understanding of trade-offs associated with collections of surrogate models operating in tandem in digital twins, the effects of multiphysics coupling on surrogate model accuracy, performance in high-dimensional settings, surrogate model VVUQ—especially in extrapolatory regimes—and, for data-driven surrogates, costs of generating training data and learning.

Surrogate Modeling for High-Dimensional, Complex Multidisciplinary Systems

State-of-the-art surrogate modeling has made considerable progress for simpler systems but remains an open challenge at the level of complexity needed for digital twins. Multiple interacting disciplines and nonlinear coupling among disciplines, as needed in a digital twin, pose a particular challenge for surrogate modeling. The availability of accurate and computationally efficient surrogate models depends on the ability to identify and exploit structure that is amenable to approximation. For example, reduced-order modeling may exploit low-rank structure in a way that permits dynamics to be evolved in a low-dimensional manifold or coarse-graining of only a subset of features, while statistical data-fit methods exploit the computational efficiencies of representing complex dynamics with a surrogate input-output map, such as a Gaussian process model or deep neural network. A challenge with coupled multidisciplinary systems is that coupling is often a key driver of dynamics—that is, the essential system dynamics can change dramatically due to coupling effects.

One example of this is Earth system models that must represent the dynamics of the atmosphere, ocean, sea ice, land surface, and cryosphere, all of which interact with each other in complex, nonlinear ways that result in interactions of processes occurring across a wide range of spatial and temporal scales. The interactions involve fluxes of mass, energy (both heat and radiation), and momentum that are dependent on the states of the various system components. Yet in many cases, the surrogate models are derived for the individual model components separately, and then coupled.

Surrogate models for coupled systems—whether data-fit or reduced-order models— remain a challenge because even if the individual model components are highly accurate representations of the dynamics and processes in those components, they may lose much of their fidelity when additional degrees of freedom due to coupling with other system components are added. Another set of challenges encompass important mathematical questions around the consistency, stability, and property-preservation attributes of coupled surrogates. A further challenge is ensuring model fidelity and fitness for purpose when multiple physical processes interact.

Finding 3-7: State-of-the-art literature and practice show advances and successes in surrogate modeling for models that form one discipline or one component of a complex system, but theory and methods for surrogates of coupled multiphysics systems are less mature.

An additional further challenge in dealing with surrogate models for digital twins of complex multidisciplinary systems is that the dimensionality of the parameter spaces underlying the surrogates can become high. For example, a surrogate model of the structural health of an engineering structure (e.g., building, bridge, airplane wing) would need to be representative over many thousands of material and structural properties that capture variation over space and time. Similarly, a surrogate model of tumor evolution in a cancer patient digital twin would potentially have thousands of parameters representing patient anatomy, physiology, and mechanical properties, again capturing variation over space and time. Deep neural networks have shown promise in representing input-output maps even when the input parameter dimension is large, yet generating sufficient training data for these complex problems remains a challenge. As discussed below, notable in the literature is that many apparent successes in surrogate modeling fail to report the cost of training, either for determining parameters in a neural network or in tuning the parameters in a reduced-order model.

It also remains a challenge to quantify the degree to which surrogate predictions may generalize in a high-dimensional setting. While mathematical advances are revealing rigorous insights into high-dimensional approximation (Cohen and DeVore 2015), this work is largely for a class of problems that exhibit smooth dynamics. Work is needed to bridge the gap between rigorous theory in high-dimensional approximation and the complex models that will underlie digital twins. Another promising set of approaches uses mathematical decompositions to break a high-dimensional problem into a set of coupled smaller-dimension problems. Again, recent advances have demonstrated significant benefits, including in the digital twin setting (Sharma et al. 2018), but these approaches have largely been limited to problems within structural modeling.

Finding 3-8: Digital twins will typically entail high-dimensional parameter spaces. This poses a significant challenge to state-of-the-art surrogate modeling methods.

Another challenge associated with surrogate models in digital twins is accounting for the data and computational resources needed to develop data-driven surrogates. While the surrogate modeling community has developed several compelling approaches in recent years, analyses of the speedups associated with these approaches in many cases do not account for the time and expense associated with generating training data or using complex numerical solvers at each iteration of the training process. A careful accounting of these elements is essential to understanding the cost–benefit trade-offs associated with surrogate models in digital twins. In tandem, advances in surrogate modeling methods for handling limited training data are needed.

Finding 3-9: One of the challenges of creating surrogate models for high-dimensional parameter spaces is the cost of generating sufficient training data. Many papers in the literature fail to properly acknowledge and report the excessively high costs (in terms of data, hardware, time, and energy consumption) of training.

Conclusion 3-2: In order for surrogate modeling methods to be viable and scalable for the complex modeling situations arising in digital twins, the cost of surrogate model training, including the cost of generating the training data, must be analyzed and reported when new methods are proposed.

Finally, the committee again emphasizes the importance of VVUQ. As noted above for hybrid modeling, development of new surrogate modeling methods must incorporate VVUQ as an integral component. While data-driven surrogate modeling methods are attractive because they reduce the computational intractability of complex modeling and require limited effort to implement, important questions remain about how well they generalize or extrapolate in realms beyond the experience of their training data. This is particularly relevant in the context of digital twins, where ideally the digital twin would explore "what if" scenarios, potentially far from the domain of the available training data—that is, where the digital twin must extrapolate to previously unseen settings. While incorporating physical models, constraints, and symmetries into data-driven surrogate models may facilitate better extrapolation performance than a generic data-driven approach, there is a lack of fundamental understanding of how to select a surrogate model approach to maximize extrapolation performance beyond empirical testing. Reduced-order models are supported by literature establishing their theoretical properties and developing error estimators for some classes of systems. Extending this kind of rigorous work may enable surrogates to be used for extrapolation with guarantees of confidence.

Data Assimilation, Dynamic Updating, and Adaptation of Surrogate Models

Dynamic updating and model adaptation are central to the digital twin concept. In many cases, this updating must be done on the fly under computational and time constraints. Surrogates play a role in making this updating computationally feasible. At the same time, the surrogate models themselves must be updated—and correspondingly validated—as the digital twin virtual representation evolves.

One set of research gaps is around the role of a surrogate model in accelerating digital twin state estimation (data assimilation) and parameter estimation (inverse problem). Challenges surrounding data assimilation and model updating in general are discussed further in Chapter 5. While data assimilation with surrogate models has been considered in some settings, it has not been extended to the scale and complexity required for the digital twin setting. Research at the intersection of data assimilation and surrogate models is an important gap. For example, data assimilation attempts to produce a state estimate by optimally combining observations and model simulation in a probabilistic framework. For data assimilation with a surrogate model to be effective, the surrogate model needs to simulate the state of the physical system accurately enough so that the difference between simulated and observed states is small. Often the parameters in the surrogate model itself are informed by data assimilation, which can introduce circularity of error propagation.

A second set of gaps is around adaptation of the surrogate models themselves. Data-fit surrogate models and reduced-order models can be updated as more data become available—an essential feature for digital twins. Entailing multiphysics coupling and high-dimensional parameter spaces as discussed above, the digital twin setting provides a particular challenge to achieving adaptation under computational constraints. Furthermore, the adaptation of a surrogate model will require an associated continual VVUQ workflow—which again must be conducted under computational constraints—so that the adapted surrogate may be used with confidence in the virtual-to-physical digital twin decision-making tasks.

KEY GAPS, NEEDS, AND OPPORTUNITIES

In Table 3-1, the committee highlights key gaps, needs, and opportunities for realizing the virtual representation of a digital twin. This is not meant to be an exhaustive list of all opportunities presented in the chapter. For the purposes of this report, prioritization of a gap is indicated by 1 or 2. While the committee believes all of the gaps listed are of high priority, gaps marked 1 may benefit from initial investment before moving on to gaps marked with a priority of 2.

TABLE 3-1 Key Gaps, Needs, and Opportunities for Realizing the Virtual Representation of a Digital Twin

Maturity	Priority
Early and Preliminary Stages	
Increasing the available computing resources for digital twin development and use is a necessary element for closing the gap between simulated and actionable scales and for engaging a broader academic community in digital twin research. Certain domains and sectors have had more success, such as engineering physics and sciences, as well as national labs.	1
Model validation and calibration for hybrid modeling are difficult given the diverse nature of the involved data and mechanistic models and their underlying assumptions. Validating data-driven models relies on sufficient and representative validation data for training, evaluation of model accuracy, and evaluation of model generalizability to new data. On the other hand, mechanistic-driven models rely on calibration and parameter estimation to accurately reproduce against experimental and independent data. Harmonizing the validation and calibration processes for these hybrid models is a gap that must be overcome to ensure the required accuracy and reliability.	2
Uncertainty quantification, explainability, and interpretability are often difficult for hybrid modeling as these systems must account for uncertainties arising from both the data-driven and mechanistic-driven components of the model as well as their interplay. Particular areas of need include uncertainty quantification for dynamically updated hybrid models, for hybrid models in extrapolative regimes, and for rare or extreme events. Warnings for extrapolations are particularly important in digital twins of critical systems.	1

continued

TABLE 3-1 Continued

Maturity	Priority
Using hybrid models can be computationally demanding and require diverse skill sets. Striking a balance between techniques that manage the computational complexity of mechanistic models and techniques used in data-driven models is essential but requires that researchers have fluency in the various approaches. Resolving how to achieve this combination and balance at a feasible and affordable level is a gap that needs to be addressed. Additionally, the model will need to be monitored and updated as time and conditions change, requiring the development of model management capabilities.	2
Uncertainty quantification is often used to calibrate component models and evaluate their fitness for purpose. However, there is a gap in understanding the sources of and quantifying the uncertainty in digital twins of coupled complex systems, due to error propagation and nonlinear feedback between the components/subsystems.	1
Interoperability is a challenge when integrating component and subsystem digital twins. There is a gap in understanding approaches to enhancing semantic and syntactic interoperability between digital twin models and reconciling assumptions made between models.	1
Coupled multiphysics systems pose particular challenges to surrogate modeling approaches that are not addressed by state-of-the-art methodology. There is a gap between the complexity of problems for which mathematical theory and scalable algorithms exist for surrogate modeling and the class of problems that underlies high-impact applications of digital twins.	2
Surrogate modeling methods that are effective when training data are limited are a gap in the state of the art. An additional related gap is methods for accounting for extrapolation.	2
There is a gap in the theory and methods to achieve dynamic adaptation of surrogate models under computational constraints, along with continual verification, validation, and uncertainty quantification (VVUQ) to assure and ensure surrogate model accuracy.	2
The consequences of the choice of prior distributions on Bayesian solutions for parameter estimation and VVUQ in general needs to be explored for both big and small data scenarios.	2
Some Research Base Exists But Additional Investment Required	
Mathematical and algorithmic advances in data-driven modeling and multiscale physics-based modeling are necessary elements for closing the gap between simulated and actionable scales. Reductions in computational and data requirements achieved through algorithmic advances are an important complement to increased computing resources. Certain domains and sectors have had more success, such as engineering and the atmospheric and climate sciences.	1

TABLE 3-1 Continued

Maturity	Priority
Combining data-driven models with mechanistic models requires effective coupling techniques to facilitate the flow of information (data, variables, etc.) between the models while understanding the inherent constraints and assumptions of each model. Coupling and model integration are complex and require harmonizing different scales, assumptions, constraints, and equations, and understanding their implications on the uncertainty associated with the outcome.	2
For the high-dimensional approximation methods where rigorous theory exists, there is a gap between the class of problems that have been considered and the class of problems that underlies high-impact applications of digital twins.	2
Research Base Exists with Opportunities to Advance Digital Twins	
The variety and coupled nature of models employed in digital twins pose particular challenges to assessing model fitness for purpose. There is a gap between the complexity of problems for which mathematical theory and scalable algorithms for error estimation exist and the class of problems that underlies high-impact applications of digital twins.	2

REFERENCES

Ainsworth, M., and J.T. Oden. 1997. "A Posteriori Error Estimation in Finite Element Analysis." *Computer Methods in Applied Mechanics and Engineering* 142(1–2):1–88.

Alber, M., A. Buganza Tepole, W.R. Cannon, S. De, S. Dura-Bernal, K. Garikipati, G. Karniadakis, et al. 2019. "Integrating Machine Learning and Multiscale Modeling—Perspectives, Challenges, and Opportunities in the Biological, Biomedical, and Behavioral Sciences." *npj Digital Medicine* 2(1):115.

Alonso, J., S. Hahn, F. Ham, M. Herrmann, G. Iaccarino, G. Kalitzin, P. LeGresley, et al. 2006. "CHIMPS: A High-Performance Scalable Module for Multiphysics Simulations." Aerospace Research Council, pp. 1–28.

Aviation Week Network. 2019. "Emirates Cuts Unscheduled Engine Removals by One-Third." *Aviation Week Network*, May 16. https://aviationweek.com/special-topics/optimizing-engines-through-lifecycle/emirates-cuts-unscheduled-engine-removals-one.

Bauer, P., B. Stevens, and W. Hazeleger. 2021. "A Digital Twin of Earth for the Green Transition." *Nature Climate Change* 11(2):80–83.

Cohen, A., and R. DeVore. 2015. "Approximation of High-Dimensional Parametric PDEs." *Acta Numerica* 24:1–159.

Deshmukh, D. 2022. "Aviation Fleet Management: Transformation Through AI/Machine Learning." *Global Gas Turbine News* 62(2):56–57.

Ferrari, A. 2023. "Building Robust Digital Twins." Presentation to the Committee on Foundational Research Gaps and Future Directions for Digital Twins. April 24. Washington, DC.

Grieves, M. 2014. "Digital Twin: Manufacturing Excellence Through Virtual Factory Replication." White paper. Michael W. Grieves LLC.

Hartmann, D., M. Herz, and U. Wever. 2018. "Model Order Reduction a Key Technology for Digital Twins." Pp. 167–179 in *Reduced-Order Modeling (ROM) for Simulation and Optimization: Powerful Algorithms as Key Enablers for Scientific Computing.* Cham, Germany: Springer International Publishing.

NASEM (National Academies of Sciences, Engineering, and Medicine). 2023a. *Opportunities and Challenges for Digital Twins in Atmospheric and Climate Sciences: Proceedings of a Workshop—in Brief.* Washington, DC: The National Academies Press.

NASEM. 2023b. *Opportunities and Challenges for Digital Twins in Biomedical Research: Proceedings of a Workshop—in Brief*. Washington, DC: The National Academies Press.

NASEM. 2023c. *Opportunities and Challenges for Digital Twins in Engineering: Proceedings of a Workshop—in Brief*. Washington, DC: The National Academies Press.

Sharma, P., D. Knezevic, P. Huynh, and G. Malinowski. 2018. "RB-FEA Based Digital Twin for Structural Integrity Assessment of Offshore Structures." Offshore Technology Conference. April 30–May 3. Houston, TX.

Sieger, M. 2019. "Getting More Air Time: This Software Helps Emirates Keep Its Planes Up and Running." *General Electric*, February 20. https://www.ge.com/news/reports/getting-air-time-software-helps-emirates-keep-planes-running.

Yankeelov, T. 2023. "Digital Twins in Oncology." Presentation to the Workshop on Opportunities and Challenges for Digital Twins in Biomedical Sciences. January 30. Washington, DC.

4

The Physical Counterpart: Foundational Research Needs and Opportunities

Digital twins rely on observation of the physical counterpart in conjunction with modeling to inform the virtual representation (as discussed in Chapter 3). In many applications, these data will be multimodal, coming from disparate sources, and of varying quality. Only when high-quality, integrated data are combined with advanced modeling approaches can the synergistic strengths of data- and model-driven digital twins be realized. This chapter addresses data acquisition and data integration for digital twins. While significant literature has been devoted to the science and best practices around gathering and preparing data for use, this chapter focuses on the most important gaps and opportunities that are crucial for robust digital twins.

DATA ACQUISITION FOR DIGITAL TWINS

Data collection for digital twins is a continual process that plays a critical role in the development, refinement, and validation of the models that comprise the virtual representation.

The Challenges Surrounding Data Acquisition for Digital Twins

Undersampling in complex systems with large spatiotemporal variability is a significant challenge for acquiring the data needed to characterize and quantify the dynamic physical and biological systems for digital twin development.

The complex systems that may make up the physical counterpart of a digital twin often exhibit intricate patterns, nonlinear behaviors, feedback, and emergent phenomena that require comprehensive sampling in order to develop an understanding of system behaviors. Systems with significant spatiotemporal variability may also exhibit heterogeneity because of external conditions, system dynamics, and component interactions. However, constraints in resources, time, or accessibility may hinder the gathering of data at an adequate frequency or resolution to capture the complete system dynamics. This undersampling could result in an incomplete characterization of the system and lead to overlooking critical events or significant features, thus risking the accuracy and predictive capabilities of digital twins. Moreover, undersampling introduces a level of uncertainty that could propagate through a digital twin's predictive models, potentially leading to inaccurate or misleading outcomes. Understanding and quantifying this uncertainty is vital for assessing the reliability and limitations of the digital twin, especially in safety-critical or high-stakes applications. To minimize the risk and effects of undersampling, innovative sampling approaches can be used to optimize data collection. Additionally, statistical methods and undersampling techniques may be leveraged to mitigate the effects of limited data.

Finally, data acquisition efforts are often enhanced by a collaborative and multidisciplinary approach, combining expertise in data acquisition, modeling, and system analysis, to address the task holistically and with an understanding of how the data will move through the digital twin.

Data Accuracy and Reliability

Digital twin technology relies on the accuracy and reliability of data, which requires tools and methods to ensure data quality, efficient data storage, management, and accessibility. Standards and governance policies are critical for data quality, accuracy, and integrity, and frameworks play an important role in providing standards and guidelines for data collection, management, and sharing while maintaining data security and privacy (see Box 4-1). Efficient and secure data flow is essential for the success of digital twin technology, and research is needed to develop cybersecurity measures; methods for verifying trustworthiness, reliability, and accuracy; and standard methods for data flow to ensure compatibility between systems. Maintaining confidentiality and privacy is also vital.

Data quality assurance is a subtle problem that will need to be addressed differently in different contexts. For instance, a key question is how a digital twin should handle outlier or anomalous data. In some settings, such data may be the result of sensor malfunctions and should be detected and ignored, while in other settings, outliers may correspond to rare events that are essential to create an accurate virtual representation of the physical counterpart. A key research challenge for digital twins is the development of methods for data quality assessment that ensure digital twins are robust to spurious outliers while accurately representing salient rare events. Several technical challenges must be addressed here. Anomaly detection is central to identifying potential issues with data quality. While anomaly detection has been studied by the statistics and signal processing communi-

BOX 4-1
Ethics and Privacy

When data from human subjects or sensitive systems are involved, privacy requirements may limit data type and volume as well as the types of computation that can be performed on them. For example, the Health Insurance Portability and Accountability Act[a] and the Common Rule[b] have specific guidance on what types of data de-identification processes must be followed when using data from electronic health records.[a] They describe how this type of data can be stored, transmitted, and used for secondary purposes. The General Protection Data Rule[c] contains similar guidance, but it extends well beyond electronic health records to most data containing personally identifiable data. In addition to legal requirements, ethical and institutional considerations are involved when utilizing real-world data. Increasing calls for transparency in the utilization of personal data have been made,[d] as well as for personal control of data, with several examples related to electronic health records.[e] These regulations that call for de-identification of the data do not address the elevated risks to privacy in the context of digital twins, and updates to regulations and data protection practices will need to address specific risks associated with digital twins.

Modelers need to keep this in mind when designing systems that will typically require periodic collection of data, as study protocols submitted to human subjects' protections programs (Institutional Review Boards) should explain the need for continuous updates (as in registries) as well as the potential for harm (as in interventional studies). Of note, model outputs may also be subject to privacy protections, since they may reveal patient information that could be used to harm patients directly or indirectly (e.g., by revealing a high probability of developing a specific health condition or by lowering their ranking in an organ transplantation queue).

---

[a] Department of Health and Human Services, 2013, "Modifications to the HIPAA Privacy, Security, Enforcement, and Breach Notification Rules Under the Health Information Technology for Economic and Clinical Health Act and the Genetic Information Nondiscrimination Act; Other Modifications to the HIPAA Rules," *Federal Register* 78(17):5565–5702.

[b] Department of Health and Human Services, 2018, "Revised Common Rule," 45 CFR Part 46.

[c] General Data Protection Regulation, 2016, "Regulations," *Official Journal of the European Union* 59(L119).

[d] State of California Legislative Council Bureau, 2018, "AB-375 Privacy: Personal Information: Businesses," Chapter 55, Title 1.81.5.

[e] J. Kim, H. Kim, E. Bell, T. Bath, P. Paul, A. Pham, X. Jiang, K. Zheng, and L. Ohno-Machado, 2019, "Patient Perspectives About Decisions to Share Medical Data and Biospecimens for Research," *JAMA Network Open* 2(8):1–13.

ties, unique challenges arise in the bidirectional feedback loop between virtual and physical systems that is inherent to digital twins, including the introduction of statistical dependencies among samples; the need for real-time processing; and heterogeneous, large-scale, multiresolution data. Another core challenge is that many machine learning (ML) and artificial intelligence (AI) methods that might be used to update virtual models from new physical data focus on maximizing average-case performance—that is, they may yield large errors on rare events. Developing digital twins that do not ignore salient rare events requires rethinking loss functions and performance metrics used in data-driven contexts.

A fundamental challenge in decision-making may arise from discrepancies between the data streamed from the physical model and that which is predicted by the digital twin. In the case of an erroneous sensor on a physical model, how can a human operator trust the output of the virtual representation, given that the supporting data were, at some point, attained data from the physical counterpart? While sensors and other data collection devices have reliability ratings, additional measures such as how reliability degrades over time may need to be taken into consideration. For example, a relatively new physical sensor showing different output compared to its digital twin may point to errors in the virtual representation instead of the physical sensor. One potential cause may be that the digital twin models may not have had enough training data under diverse operating conditions that capture the changing environment of the physical counterpart.

Data quality (e.g., ensuring that the data set is accurate, complete, valid, and consistent) is another major concern for digital twins. Consider data assimilation for the artificial pancreas or closed-loop pump (insulin and glucagon). The continuous glucose monitor has an error range, as does the glucometer check, which itself is dependent on compliance from the human user (e.g., washing hands before the glucose check). Data assimilation techniques for digital twins must be able to handle challenges with multiple inputs from the glucose monitor and the glucometer, especially if they provide very different glucose levels, differ in units from different countries (e.g., mmol/L or mg/dL), or lack regular calibration of the glucometer. Assessing and documenting data quality, including completeness and measures taken to curate the data, tools used at each step of the way, and benchmarks against which any model is evaluated, are integral parts of developing and maintaining a library of reproducible models that can be embedded in a digital twin system.

> Finding 4-1: Documenting data quality and the metadata that reflect the data provenance is critical.

Without clear guidelines for defining the objectives and use cases of digital twin technology, it can be challenging to identify critical components that significantly impact the physical system's performance (VanDerHorn and Mahadevan

2021). The absence of standardized quality assurance frameworks makes it difficult to compare and validate results across different organizations and systems.

Finding 4-2: The absence of standardized quality assurance frameworks makes it difficult to compare and validate results across different organizations and systems. This is important for cybersecurity and information and decision sciences. Integrating data from various sources, including Internet of Things devices, sensors, and historical data, can be challenging due to differences in data format, quality, and structure.

Considerations for Sensors

Sensors provide timely data on the condition of the physical counterpart. Improvements in sensor integrity, performance, and reliability will all play a crucial role in advancing the reliability of digital twin technology; this requires research into sensor calibration, performance, maintenance, and fusion methods. Detecting and mitigating adversarial attacks on sensors, such as tampering or false data injection, is essential for preserving system integrity and prediction fidelity. Finally, multimodal sensors that combine multiple sensing technologies may enhance the accuracy and reliability of data collection. Data integration is explored further in the next section. A related set of research questions around optimal sensor placement, sensor steering, and sensor dynamic scheduling is discussed in Chapter 6.

DATA INTEGRATION FOR DIGITAL TWINS

Increased access to diverse and dynamic streams of data from sensors and instruments can inform decision-making and improve model reliability and robustness. The digital twin of a complex physical system often gets data in different formats from multiple sources with different levels of verification and validation (e.g., visual inspection, record of repairs and overhauls, and quantitative sensor data from a limited number of locations). Integrating data from various sources—including Internet of Things devices, sensors, and historical data—can be challenging due to differences in data format, quality, and structure. Data interoperability (i.e., the ability for two or more systems to exchange and use information from other systems) and integration are important considerations for digital twins, but current efforts toward semantic integration are not scalable. Adequate metadata are critical to enabling data interoperability, harmonization, and integration, as well as informing appropriate use (Chung and Jaffray 2021). The transmission and level of key information needed and how to incorporate it in the digital twin are not well understood, and efforts to standardize metadata exist but are not yet sufficient for the needs of digital twins. Developers and end users would benefit from collaboratively addressing the needed type and format of data prior to deployment.

Handling Large Amounts of Data

In some applications, data may be streaming at full four-dimensional resolution and coupled with applications on the fly. This produces significantly large amounts of data for processing. Due to the large and streaming nature of some data sets, all operations must be running in continuous or on-demand modes (e.g., ML models need to be trained and applied on the fly, applications must operate in fully immersive data spaces, and data assimilation and data handling architecture must be scalable). Specific challenges around data assimilation and the associated verification, validation, and uncertainty quantification efforts are discussed further in Chapter 5. Historically, data assimilation methods have been model-based and developed independently from data-driven ML models. In the context of digital twins, however, these two paradigms will require integration. For instance, ML methods used within digital twins need to be optimized to facilitate data assimilation with large-scale streaming data, and data assimilation methods that leverage ML models, architectures, and computational frameworks need to be developed.

The scalability of data storage, movement, and management solutions becomes an issue as the amount of data collected from digital twin systems increases. In some settings, the digital twin will face computational resource constraints (e.g., as a result of power constraints); in such cases, low-power ML and data assimilation methods are required. Approaches based on subsampling data (i.e., only using a subset of the available data to update the digital twin's virtual models) necessitate statistical and ML methods that operate reliably and robustly with limited data. Foundational research on the sample complexity of ML methods as well as pretrained and foundational models that only require limited data for fine tuning are essential to this endeavor. Additional approaches requiring further research and development include model compression, which facilitates the efficient evaluation of deployed models; dimensionality reduction (particularly in dynamic environments); and low-power hardware or firmware deployments of ML and data assimilation tools.

In addition, when streaming data are being collected and assimilated continuously, models must be updated incrementally. Online and incremental learning methods play an important role here. A core challenge is setting the learning rate in these models. The learning rate controls to what extent the model retains its memory of past system states as opposed to adapting to new data. This rate as well as other model hyperparameters must be set and tuned on the fly, in contrast to the standard paradigm of offline tuning using holdout data from the same distribution as training data. Methods for adaptively setting a learning rate, so that it is low enough to provide robustness to noisy and other data errors when the underlying state is slowly varying yet can be increased when the state changes sharply (e.g., in hybrid or switched dynamical systems), are a critical research challenge for digital twins. Finally, note that the data quality challenges outlined above are present in the large-scale streaming data setting as well, making the challenge of adaptive model training in the presence of anomalies and outliers that may correspond to either sensor failures or salient rare events particularly challenging.

Data Fusion and Synchronization

Digital twins can integrate data from different data streams, which provides a means to address missing data or data sparsity, but there are specific concerns regarding data synchronization (e.g., across scales) and data interoperability. For example, the heterogeneity of data sources (e.g., data from diverse sensor systems) can present challenges for data assimilation in digital twins. Specific challenges include the need to estimate the impact of missing data as well as the need to integrate data uncertainties and errors in future workflows. The integration of heterogeneous data requires macro to micro levels of statistical synthesis that span multiple levels, scales, and fidelities. Moreover, approaches must be able to handle mismatched digital representations. Recent efforts in the ML community on multiview learning and joint representation learning of data from disparate sources (e.g., learning a joint representation space for images and their text captions, facilitating the automatic captioning of new images) provide a collection of tools for building models based on disparate data sources.

For example, in tumor detection using magnetic resonance imaging (MRI), results depend on the radiologist identifying the tumor and measuring the linear diameter manually (which is susceptible to inter- and intra-observer variability). There are efforts to automate the detection, segmentation, and/or measurement of tumors (e.g., using AI and ML approaches), but these are still vulnerable to upstream variability in image acquisition (e.g., a very small 2 mm tumor may be detected on a high-quality MRI but may not be visible on a poorer quality machine). Assimilating serial tumor measurement data is a complex challenge due to patients being scanned in different scanners with different protocols over time.

Data fusion and synchronization are further exacerbated by disparate sampling rates, complete or partial duplication of records, and different data collection contexts, which may result in seemingly contradictory data. The degree to which data collection is done in real time (or near real time) is dependent on the intended purpose of the digital twin system as well as available resources. For example, an ambulatory care system has sporadic electronic health record data, while intensive care unit sensor data are acquired at a much faster sampling rate. Additionally, in some systems, data imputation to mitigate effects of missing data will also require the development of imputation models learned from data.

Lack of standardization creates interoperability issues while integrating data from different sources.

Conclusion 4-1: The lack of adopted standards in data generation hinders the interoperability of data required for digital twins. Fundamental challenges include aggregating uncertainty across different data modalities and scales as well as addressing missing data. Strategies for data sharing and collaboration must address challenges such as data ownership and intellectual property issues while maintaining data security and privacy.

Challenges with Data Access and Collaboration

Digital twins are an inherently multidisciplinary and collaborative effort. Data from multiple stakeholders may be integrated and/or shared across communities. Strategies for data collaboration must address challenges such as data ownership, responsibility, and intellectual property issues prior to data usage and digital twin deployment.

Some of these challenges can be seen in Earth science research, which has been integrating data from multiple sources for decades. Since the late 1970s, Earth observing satellites have been taking measurements that provide a nearly simultaneous global estimate of the state of the Earth system. When combined through data assimilation with in situ measurements from a variety of platforms (e.g., surface stations, ships, aircraft, and balloons), they provide global initial conditions for a numerical model to produce forecasts and also provide a basis for development and improvement of models (Ackerman et al. 2019; Balsamo et al. 2018; Fu et al. 2019; Ghil et al. 1979). The combination of general circulation models of the atmosphere, coupled models of the ocean–atmosphere system, and Earth system models that include biogeochemical models of the carbon cycle together with global, synoptic observations and a data assimilation method represent a digital twin of the Earth system that can be used to make weather forecasts and simulate climate variability and change. Numerical weather prediction systems are also used to assess the relative value of different observing systems and individual observing stations (Gelaro and Zhu 2009).

KEY GAPS, NEEDS, AND OPPORTUNITIES

In Table 4-1, the committee highlights key gaps, needs, and opportunities for managing the physical counterpart of a digital twin. There are many gaps, needs, and opportunities associated with data management more broadly; here the committee focuses on those for which digital twins bring unique challenges. This is not meant to be an exhaustive list of all opportunities presented in the chapter. For the purposes of this report, prioritization of a gap is indicated by 1 or 2. While the committee believes all of the gaps listed are of high priority, gaps

TABLE 4-1 Key Gaps, Needs, and Opportunities for Managing the Physical Counterpart of a Digital Twin

Maturity	Priority
Early and Preliminary Stages	
Standards to facilitate interoperability of data and models for digital twins (e.g., by regulatory bodies) are lacking.	1
Undersampling in complex systems with large spatiotemporal variability is a significant challenge for acquiring the data needed for digital twin development. This undersampling could result in an incomplete characterization of the system and lead to overlooking critical events or significant features. It also introduces uncertainty that could propagate through the digital twin's predictive models, potentially leading to inaccurate or misleading outcomes. Understanding and quantifying this uncertainty is vital for assessing the reliability and limitations of the digital twin, especially in safety-critical or high-stakes applications.	2
Data imputation approaches for high volume and multimodal data are needed.	2
Some Research Base Exists But Additional Investment Required	
Tools are needed for data and metadata handling and management to ensure that data and metadata are gathered, stored, and processed efficiently.	1
There is a gap in the mathematical tools available for assessing data quality, determining appropriate utilization of all available information, understanding how data quality affects the performance of digital twin systems, and guiding the choice of an appropriate algorithm.	2

marked 1 may benefit from initial investment before moving on to gaps marked with a priority of 2.

REFERENCES

Ackerman, S.A, S. Platnick, P.K. Bhartia, B. Duncan, T. L'Ecuyer, A. Heidinger, G.J. Skofronick, N. Loeb, T. Schmit, and N. Smith. 2019. "Satellites See the World's Atmosphere." *Meteorological Monographs* 59(1):1–53.

Balsamo, G., A.A. Parareda, C. Albergel, C. Arduini, A. Beljaars, J. Bidlot, E. Blyth, et al. 2018. "Satellite and In Situ Observations for Advancing Global Earth Surface Modelling: A Review." *Remote Sensing* 10(12):2038.

Chung, C., and D. Jaffray. 2021. "Cancer Needs a Robust 'Metadata Supply Chain' to Realize the Promise of Artificial Intelligence." *American Association for Cancer Research* 81(23):5810–5812.

Fu, L.L., T. Lee, W.T. Liu, and R. Kwok. 2019. "50 Years of Satellite Remote Sensing of the Ocean." *Meteorological Monographs* 59(1):1–46.

Gelaro, R., and Y. Zhu. 2009. "Examination of Observation Impacts Derived from Observing System Experiments (OSEs) and Adjoint Models." *Tellus A: Dynamic Meteorology and Oceanography* 61(2):179–193.

Ghil, M., M. Halem, and R. Atlas. 1979. "Time-Continuous Assimilation of Remote-Sounding Data and Its Effect on Weather Forecasting." *Monthly Weather Review* 107(2):140–171.

VanDerHorn, E., and S. Mahadevan. 2021. "Digital Twin: Generalization, Characterization and Implementation." *Decision Support Systems* 145:113524.

5

Feedback Flow from Physical to Virtual: Foundational Research Needs and Opportunities

In the digital twin feedback flow from physical to virtual, inverse problem methodologies and data assimilation are required for combining physical observations and virtual models in a rigorous, systematic, and scalable way. This chapter addresses specific challenges for digital twins including calibration and updating on actionable time scales. These challenges represent foundational gaps in inverse problem and data assimilation theory, methodology, and computational approaches.

INVERSE PROBLEMS AND DIGITAL TWIN CALIBRATION

Digital twin calibration is the process of estimating numerical model parameters for individualized digital twin virtual representations. This task of estimating numerical model parameters and states that are not directly observable can be posed mathematically as an inverse problem, but the problem may be ill posed. Bayesian approaches can be used to incorporate expert knowledge that constrains solutions and predictions. It must be noted, however, that for some settings, specification of prior distributions can greatly impact the inferences that a digital twin is meant to provide—for better or for worse. Digital twins present specific challenges to Bayesian approaches, including the need for good priors that capture tails of distributions, the need to incorporate model errors and updates, and the need for robust and scalable methods under uncertainty and for high-consequence decisions. This presents a new class of open problems in the realm of inverse problems for large-scale complex systems.

Parameter Estimation and Regularization for Digital Twin Calibration

The process of estimating numerical model parameters from data is an ill-posed problem, whereby the solution may not exist, may not be unique, or may not depend continuously on the data. The first two conditions are related to *identifiability* of solutions. The third condition is related to the *stability* of the problem; in some cases, small errors in the data may result in large errors in the reconstructed parameters. Bayesian regularization, in which priors are encoded using probability distribution functions, can be used to handle missing information, ill-posedness, and uncertainty. A specific challenge for digital twins is that standard priors—such as those based on simple Gaussian assumptions—may not be informative and representative for making high-stakes decisions. Also, due to the continuous feedback loop, updated models need to be included on the fly (without restarting from scratch). Moreover, the prior for one problem may be taken as the posterior from a previous problem, so it is important to assign probabilities to data and priors in a rigorous way such that the posterior probability is consistent when using a Bayesian framework.

Approaches to learn priors through existing data (e.g., machine learning–informed bias correction) can work well in data-rich environments but may not accurately represent or predict extreme events because of limited relevant training data. Bayesian formulations require priors for the unknown parameters, which may depend on expensive-to-tune hyperparameters. Data-driven regularization approaches that incorporate more realistic priors are necessary for digital twins.

Optimization of Numerical Model Parameters Under Uncertainty

Another key challenge is to perform optimization of numerical model parameters (and any additional hyperparameters) under uncertainty—any computational model must be calibrated to meet its requirements and be fit for purpose. In general, optimization under uncertainty is challenging because the cost functions are stochastic and must be able to incorporate different types of uncertainty and missing information. Bayesian optimization and stochastic optimization approaches (e.g., online learning) can be used, and some fundamental challenges—such as obtaining sensitivity information from legacy code with missing adjoints—are discussed in Chapter 6.

These challenges are compounded for digital twin model calibration, especially when models are needed at multiple resolutions. Methods are needed for fast sampling of parametric and structural uncertainty. For digital twins to support high-consequence decisions, methods may need to be tuned to risk and extreme events, accounting for worst-case scenarios. Risk-adaptive loss functions and data-informed prior distribution functions for capturing extreme events and for incorporating risk during inversion merit further exploration. Non-differentiability also becomes a significant concern as mathematical models may demonstrate discontinuous behavior or numerical artifacts may result in models that appear non-differentiable. Moreover, models may even be chaotic, which can be intractable for adjoint and tangent linear models. Standard loss functions, such as the least-squares loss, are not able to model chaotic behavior in the data (Royset 2023) and are not able to represent complex statistical distributions of model errors that arise from issues such as using a reduced or low-fidelity digital-forward model. Robust and stable optimization techniques (beyond gradient-based methods) to handle new loss functions and to address high displacements (e.g., the upper tail of a distribution) that are not captured using only the mean and standard deviation are needed.

DATA ASSIMILATION AND DIGITAL TWIN UPDATING

Data assimilation tools have been used heavily in numerical weather forecasting, and they can be critical for digital twins broadly, including to improve model states based on current observations. Still, there is more to be exploited in the bidirectional feedback flow between physical and virtual beyond standard data assimilation (Blair 2021).

First, existing data assimilation methods rely heavily on assumptions of high-fidelity models. However, due to the continual and dynamic nature of digital twins, the validity of a model's assumptions—and thus the model's fidelity—may evolve over time, especially as the physical counterpart undergoes significant shifts in condition and properties. A second challenge is the need to perform uncertainty quantification for high-consequence decisions on actionable time scales. This becomes particularly challenging for large-scale complex systems with high-dimensional parameter and state spaces. Direct simulations and inversions (e.g., in the case of variational methods) needed for data assimilation are no longer feasible. Third, with different digital technologies providing data at unprecedented rates, there are few mechanisms for integrating artificial intelligence, machine learning, and data science tools for updating digital twins.

Digital Twin Demands for Continual Updates

Digital twins require continual feedback from the physical to virtual, often using partial and noisy observations. Updates to the twin should be incorporated in a timely way (oftentimes immediately), so that the updated digital twin may be used for further forecasting, prediction, and guidance on where to obtain new data. These updates may be initiated when something in the physical counterpart evolves or in response to changes in the virtual representation, such as improved model parameters, a higher-fidelity model that incorporates new physical understanding, or improvements in scale/resolution. Due to the continual nature of digital twins as well as the presence of errors and noise in the models, the observations, and the initial conditions, sequential data assimilation approaches (e.g., particle-based approaches and ensemble Kalman filters) are the natural choice for state and parameter estimation. However, these probabilistic approaches have some disadvantages compared to variational approaches, such as sampling errors, rank deficiency, and inconsistent assimilation of asynchronous observations.

Data assimilation techniques need to be adapted for continuous streams of data from different sources and need to interface with numerical models with potentially varying levels of uncertainty. These methods need to be able to infer system state under uncertainty when a system is evolving and be able to integrate model updates efficiently. Moreover, navigating discrepancies between predictions and observed data requires the development of tools for model update documentation and hierarchy tracking.

Conclusion 5-1: Data assimilation and model updating play central roles in the physical-to-virtual flow of a digital twin. Data assimilation techniques are needed for data streams from different sources and for numerical models with varying levels of uncertainty. A successful digital twin will require the continuous assessment of models. Traceability of model hierarchies and reproducibility of results are not fully considered in existing data assimilation approaches.

Digital Twin Demands for Actionable Time Scales

Most literature focuses on offline data assimilation, but the assimilation of real-time sensor data for digital twins to be used on actionable time scales will require advancements in data assimilation methods and tight coupling with the control or decision-support task at hand (see Chapter 6).

For example, the vast, global observing system of the Earth's atmosphere and numerical models of its dynamics and processes are combined in a data assimilation framework to create initial conditions for weather forecasts. In order for a weather forecast to have value, it must be delivered within a short interval of real time. This requires a huge computational and communications apparatus of gathering, ingesting, processing, and assimilating global observations within a window of a few hours. High-performance computing implementations of state-of-the-art data assimilation codes and new data assimilation approaches that can exploit effective dimensionality within an optimization/outer-loop approach for obtaining optimal solutions (e.g., latent data assimilation to reduce the dimensionality of the data) are needed.

High-Consequence Decisions Demand Large-Scale Uncertainty Quantification

Data assimilation provides a framework for combining model-based predictions and their uncertainties with observations, but it lacks the decision-making interface—including measures of risk—needed for digital twins. Bayesian estimation and inverse modeling provide the mathematical tools for quantifying uncertainty about a system. Given data, Bayesian parameter estimation can be used to select the best model and to infer posterior probability distributions for numerical model parameters. Forward propagation of these distributions then leads to a posterior prediction, in which the digital twin aids decision-making by providing an estimate of the prediction quantities of interest and their uncertainties. This process provides predictions and credible intervals for quantities of interest but relies heavily on prior assumptions and risk-informed likelihoods, as well as advanced computational techniques such as Gaussian process emulators for integrating various sources of uncertainty. For solving the Bayesian inference problem, sampling approaches such as Markov chain Monte Carlo are prohibitive because of the many thousands of forward-problem solves (i.e., model simulations) that would be needed. Machine learning has the potential to support uncertainty quantification through approaches such as diffusion models or other generative artificial intelligence methods that can capture uncertainties, but the lack of theory and the need for large ensembles and data sets provide additional challenges. Increasing computational capacity alone will not address these issues.

Large Parameter Spaces and Data-Rich Scenarios

For many digital twins, the sheer number of numerical model parameters that need to be estimated and updated can present computational issues of tractability and identifiability. For example, a climate model may have hundreds of millions of spatial degrees of freedom. Performing data assimilation and optimization under uncertainty for such large-scale complex systems is not feasible. Strategies include reducing the dimensionality of the numerical model parameters via surrogate models (see Chapter 3), imposing structure or more informative priors (e.g., using Bayesian neural networks or sparsity-promoting regularizers), and developing goal-oriented approaches for problems where quantities of interest from predictions can be identified and estimated directly from the data. Goal-oriented approaches for optimal design, control, and decision support are addressed in Chapter 6.

For data-rich scenarios, there are fundamental challenges related to the integration of massive amounts of observational data being collected. For example, novel atmospheric observational platforms (e.g., smart devices that can sense atmospheric properties like temperature) provide a diversity of observational frequency, density, and error characteristics. This provides an opportunity for more rapid and timely updating of the state of the atmosphere in the digital twin, but it also represents a challenge for existing data assimilation techniques that are not able to utilize all information from various types of instrumentations.

Conclusion 5-2: Data assimilation alone lacks the learning ability needed for a digital twin. The integration of data science with tools for digital twins (including inverse problems and data assimilation) will provide opportunities to extract new insights from data.

KEY GAPS, NEEDS, AND OPPORTUNITIES

In Table 5-1, the committee highlights key gaps, needs, and opportunities for enabling the feedback flow from the physical counterpart to the virtual representation of a digital twin. This is not meant to be an exhaustive list of all opportunities presented in the chapter. For the purposes of this report, prioritization of a gap is indicated by 1 or 2. While the committee believes all of the gaps listed are of high priority, gaps marked 1 may benefit from initial investment before moving on to gaps marked with a priority of 2.

TABLE 5-1 Key Gaps, Needs, and Opportunities for Enabling the Feedback Flow from the Physical Counterpart to the Virtual Representation of a Digital Twin

Maturity	Priority
Early and Preliminary Stages	
Tools for tracking model and related data provenance (i.e., maintaining a history of model updates and tracking model hierarchies) to handle scenarios where predictions do not agree with observed data are limited. Certain domains and sectors have had more success, such as the climate and atmospheric sciences.	1
New uncertainty quantification methods for large-scale problems that can capture extreme behavior and provide reliable uncertainty and risk analysis are needed. New data assimilation methods that can handle more channels of data and data coming from multiple sources at different scales with different levels of uncertainty are also needed.	1
Some Research Base Exists But Additional Investment Required	
Risk-adaptive loss functions and data-informed prior distribution functions for capturing extreme events and for incorporating risk during inversion are needed. Also needed are robust and stable optimization techniques (beyond gradient-based methods) to handle new loss functions and to address high displacements (e.g., the upper tail of a distribution) that are not captured using only the mean and standard deviation.	1
High-performance computing implementations of state-of-the-art data assimilation codes (ranging from high-dimensional particle filters to well-studied ensemble Kalman filters, or emulators) and new data assimilation approaches that can exploit effective dimensionality within an optimization/outer-loop approach for obtaining optimal solutions (e.g., latent data assimilation to reduce the dimensionality of the data) are needed.	2

continued

TABLE 5-1 Continued

Maturity	Priority
Machine learning has the potential to support uncertainty quantification through approaches such as diffusion models or other generative artificial intelligence methods that can capture uncertainties, but the lack of theory and the need for large ensembles and data sets provides additional challenges.	2
Standards and governance policies are critical for data quality, accuracy, security, and integrity, and frameworks play an important role in providing standards and guidelines for data collection, management, and sharing while maintaining data security and privacy.	1
Research Base Exists with Opportunities to Advance Digital Twins	
New approaches that incorporate more realistic prior distributions or data-driven regularization are needed. Since uncertainty quantification is often necessary, fast Bayesian methods will need to be developed to make solutions operationally practical.	2

REFERENCES

Blair, G.S. 2021. "Digital Twins of the Natural Environment." *Patterns* 2(10):1–3.

Royset, J.O. 2023. "Risk-Adaptive Decision-Making and Learning." Presentation to the Committee on Foundational Research Gaps and Future Directions for Digital Twins. February 13. Washington, DC.

6

Feedback Flow from Virtual to Physical: Foundational Research Needs and Opportunities

On the virtual-to-physical flowpath, the digital twin is used to drive changes in the physical counterpart itself or in the sensor and observing systems associated with the physical counterpart. This chapter identifies foundational research needs and opportunities associated with the use of digital twins for automated decision-making tasks such as control, optimization, and sensor steering. This chapter also discusses the roles of digital twins for providing decision support to a human decision-maker and for decision tasks that are shared jointly within a human–agent team. The chapter concludes with a discussion of the ethical and social implications of the use of digital twins in decision-making.

PREDICTION, CONTROL, STEERING, AND DECISION UNDER UNCERTAINTY

Just as there is a broad range of model types and data that may compose a digital twin depending on the particular domain and use case, there is an equally broad range of prediction and decision tasks that a digital twin may be called on to execute and/or support. This section focuses on tasks that manifest mathematically as control and optimization problems. Examples include automated control of an engineering system, optimized treatment regimens and treatment response assessments (e.g., diagnostic imaging or laboratory tests) recommended to a human medical decision-maker, optimized sensor locations deployed over an environmental area, automated dynamic sensor steering, and many more (see Box 6-1).

When it comes to these control and optimization tasks, a digital twin has unique features that challenge existing methods and expose foundational research

BOX 6-1
Decision-Making Examples

Drug Discovery

Drug discovery often entails screening multiple candidate compounds to evaluate how well they fit a target binding site. Exhaustively testing each candidate compound can be an inefficient use of limited resources. A digital twin of molecular binding and drug synthesis may help guide the search for new pharmaceuticals. For example, decision-making techniques from contextual bandits or Bayesian optimization may provide real-time guidance for which compounds to prioritize for screening and give principled mechanisms for incorporating new data as they are collected during the screening process.

Contaminant Assessment and Control

Subsurface contaminants can threaten water supplies. It is challenging to track and control the contaminant due to the unknown properties (e.g., permeability) of the subsurface, the unknown state of the contaminant concentration over space and time, and a limited ability to observe contaminant concentrations. A digital twin of a subsurface region can guide decision-making on both sensing decisions and contaminant control decisions: where to drill observation wells, where to drill control wells, and what are optimal pumping/injection profiles at control wells.

Asset Performance Management

Equipment maintenance and replacement efforts are improved using prediction to mitigate breakdowns. Digital twins of various assets (e.g., pumps, compressors, and turbines) can serve as powerful asset management tools. In one case the digital twin of a turbine was able to remotely receive operating data and determine that the enclosure temperature was different from the predicted value in the digital model. This discrepancy led to an early warning that allowed for preemptive maintenance.[a]

Thermal Management

The internal temperature of a large motor is measured to avoid overheating. However, it may not be possible to fit temperature sensors inside the motor casing. A digital twin of a motor may predict its temperature in the absence of a sensor using power or current data in a physics-based model, for instance. Data from mechanistic simulations may be used to develop a reduced-order model for temperature. Using a digital twin for thermal management allows equipment operators to achieve higher uptimes, as compared to other methods that rely on large safety factors.[b]

Locomotive Trip Optimization

Within the locomotive industry, smart cruise control systems consider terrain, train composition, speed restrictions, and operating conditions to compute an optimal speed profile. Trip optimizers utilize digital twins of trains to autonomously

> **BOX 6-1 Continued**
>
> manage the locomotive's throttle and dynamic brakes to minimize fuel consumption at this speed profile and ensure efficient train handling. One example of locomotive speed optimization resulted in a 10% reduction in fuel consumption.[c]
>
> ---
>
> [a] GE, n.d., "Remote Monitoring, Powered by Digital Twins," https://www.ge.com/digital/industrial-managed-services-remote-monitoring-for-iiot/article/336867-early-warning-of-increased-enclosure-temperature-on-a-aeroderivative-gas-turbine, accessed September 30, 2023.
> [b] Siemens, n.d., "Virtual Sensor Opens a World of Efficiency for Large Motors," https://www.siemens.com/global/en/company/stories/research-technologies/digitaltwin/virtual-sensor-opens-a-world-of-efficiency-for-large-motors.html, accessed September 30, 2023.
> [c] Wabtec, July 27, 2020, https://www.wabteccorp.com/newsroom/press-releases/wabtec-s-trip-optimizer-system-surpasses-500-million-miles-of-operation, accessed September 30, 2023.

gaps. Similar to the modeling challenges discussed in Chapter 3, the scale and complexity of a digital twin of a multiphysics, multiscale system (or system of systems) make control and optimization computationally challenging—even if decisions can be distilled down to a small number of quantities of interest, the decision variables (i.e., controller variables, optimization design variables) and system parameters affecting the control/optimization are likely to be of high dimension. Furthermore, the importance of digital twin verification, validation, and uncertainty quantification (VVUQ) is brought to the fore when it comes to decision-making tasks, yet the requisite end-to-end quantification of uncertainty places an even greater burden on the control and optimization methods. Lastly, the highly integrated nature of a digital twin leads to a need for tight and iterative coupling between data assimilation and optimal control—possibly in real time, on deployed computing platforms, and all with quantified uncertainty. Research gaps to address these challenges span many technical areas including operations research, reinforcement learning, optimal and stochastic control, dynamical systems, partial differential equation (PDE) constrained optimization, scalable algorithms, and statistics.

Rare Events and Risk Assessment in Support of Decision-Making

In many applications, digital twins will be called on to execute or support decisions that involve the characterization of low-probability events (e.g., failure in an engineering system, adverse outcomes in a medical intervention). In these cases, using digital twins to develop the quantifiable basis for decision-making requires a careful analysis of decision metrics, with particular attention to how one quantifies risk. A gap currently exists between state-of-the-art risk quantification methods and tools used for decision-making in practical science, engineering, and medicine contexts.

For example, risk metrics such as superquantiles are widely used for decision-making in the financial industry but have seen limited adoption in engineering (Royset 2023). The barriers go beyond just awareness and are in some cases systemic; for example, for some engineering systems, metrics such as probability of failure are encoded in certification standards. The challenges in assessing risk may be compounded in the context of digital twins developed to optimize one figure of merit but then adapted for a decision-making task in which performance metrics are different. Another challenge is that many risk measures lead to a non-differentiable objective. Chance constraints are often needed to impose probabilistic constraints on system behavior, resources, etc., and these can be non-differentiable as well. Monte Carlo gradient estimation procedures may be used to handle cases of non-differentiability. In general, non-differentiability complicates the use of gradient-based optimization methods, which in turn can limit scalability, and advanced uncertainty quantification methods relying on smoothness (e.g., stochastic Galerkin, stochastic collocation) may produce large integration errors. On the other hand, Monte Carlo sampling becomes extremely inefficient, especially when dealing with low-probability events.

> Finding 6-1: There is a need for digital twins to support complex trade-offs of risk, performance, cost, and computation time in decision-making.

Sensor Steering, Optimal Experimental Design, and Active Learning

Within the realm of decision-making supported and executed by digital twins is the important class of problems that impact the data—specifically, the sensing and observing systems—of the physical counterpart. These problems may take the form of sensor placement, sensor steering, and sensor dynamic scheduling, which can be broadly characterized mathematically as optimal experimental design (OED) problems or in the data-driven literature as active learning. Just as for control problems, the needs for digital twins go beyond the capabilities of state-of-the-art methods.

Mathematically and statistically sophisticated formulations exist for OED, but there is a lack of approaches that scale to high-dimensional problems of the kinds anticipated for digital twins, while accounting for uncertainties and handling the rich modalities and complications of multiple data streams discussed in Chapter 4. Of particular relevance in the digital twin setting is the tight integration between sensing, inference, and decision-making, meaning that the OED problem cannot be considered in isolation. Rather, there is a need to integrate the OED problem with the data assimilation approaches of Chapter 5 and the control or decision-support task at hand (Ghattas 2022). That is, the sensors need to be steered in such a way as to maximize knowledge, not about the system parameters or state but about the factors feeding into the digital twin decision problem (e.g., the objective and constraints in an optimal control problem). The resulting integrated sense–assimilate–predict–control–steer cycle is challenging and cuts across several traditional areas of study but ultimately will lead to the most powerful instantiation of digital twins; therefore, scalable methods for goal-oriented sensor steering and OED over the entire sense–assimilate–predict–control–steer cycle merit further exploration.

Conclusion 6-1: There is value in digital twins that can optimally design and steer data collection, with the ultimate goal of supporting better decision-making.

Emphasizing the role of digital twins in data collection, it is crucial to recognize that they not only design and steer data gathering but also ensure the acquisition of high-fidelity, relevant, and actionable data. This capability enables more precise model training and fine-tuning, which translates to more reliable forecasts and simulations. Additionally, in the context of evolving environments and systems, digital twins play a pivotal role in adaptive data collection strategies, identifying areas that need more data and refining collection parameters dynamically. Ultimately, these capabilities, when harnessed correctly, can lead to more informed and timely decision-making, reducing risks and enhancing efficiency.

Digital Twin Demands for Real-Time Decision-Making

In several settings, control or optimization tasks may require real-time (or near-real-time) execution. The time scales that characterize real-time response may vary widely across applications, from fractions of seconds in an engineering automated control application to minutes or hours in support of a clinical decision. In many cases, achieving these actionable time scales will necessitate the use of surrogate models, as discussed in Chapter 3. These surrogate models must be predictive not just over state space but also over parameter space and decision variable space (Ghattas 2023). This places additional demands on both VVUQ and training data needs for the surrogates, compounding the challenges discussed in Chapter 3.

A mitigating factor can be to exploit the mathematical structure of the decision problem to obtain goal-oriented surrogates that are accurate with respect to the optimization/control objectives and constraints but need not reproduce the entire state space accurately. An additional challenge is that real-time digital twin computations may need to be conducted using edge computing under constraints on computational precision, power consumption, and communication. Machine learning (ML) models that can be executed rapidly are well suited to meet the computational requirements of real-time and/or in situ decision-making, but their black-box nature provides additional challenges for VVUQ and explainability. Additional work is needed to develop trusted ML and surrogate models that perform well under the computational and temporal conditions necessary for real-time decision-making with digital twins.

Finding 6-2: In many cases, trusted high-fidelity models will not meet the computational requirements to support digital twin decision-making.

Digital Twin Demands for Dynamic Adaptation in Decision-Making

A hallmark feature of digital twins is their ability to adapt to new conditions and data on the fly. In the context of decision-making, these new conditions may reflect, for instance, changes to the set of available states, the state transition probabilities themselves, or the environment. The nature of how decisions are made, particularly in automated control settings, may also necessitate dynamic adaptation. Reinforcement learning approaches address this setting, but currently there is a gap between theoretical performance guarantees for stylized settings and efficacious methods in practical domains. These challenges and gaps are exacerbated in the context of digital twins, where continual updates in response to new data require constant adaptation of decision-making methods. Safety-constrained reinforcement learning is beginning to address some of these issues for control problems in which it must be ensured that a system remains within a safe zone, particularly in the context of robotics and autonomous vehicles (Brunke et al. 2022; Isele et al. 2018).

Digital twins provide a useful mechanism for exploring the efficacy and safety of such methods. In addition, since the coupled data assimilation and optimal control problems are solved repeatedly over a moving time window, there is an opportunity to exploit dynamically adaptive optimization and control algorithms that can exploit sensitivity information to warm-start new solutions. Scalable methods to achieve dynamic adaptation in digital twin decision-making are necessary for exploiting the potential of digital twins. In the medical setting, beyond safety constraints are feasibility constraints; some treatment delivery recommendations may not be feasible (due to, for example, patient willingness or financial or system burden challenges) unless major changes are made to the entire system. For example, a recommendation of daily chemotherapy infusions could overwhelm the current system; however, home infusion capabilities may be a possible development in the future, and so feasibility constraints could also evolve over time.

Finding 6-3: Theory and methods are being developed for reinforcement learning and for dynamically adaptive optimization and control algorithms. There is an opportunity to connect these advances more strongly to the development of digital twin methodologies.

Model-Centric and Data-Centric Views of Digital Twin Decision-Making

Underlying all these research gaps and opportunities to support digital twin decision-making is the synergistic interplay between models and data. As discussed in Chapter 3, a digital twin is distinguished from traditional modeling and simulation in the way that models and data work together to drive decision-making. The relative richness or scarcity of data together with the complexity and consequence of the decision space will significantly influence the appropriateness of different approaches (Ferrari 2023).

In data-rich scenarios, digital twins offer new opportunities to develop decision-making systems without explicit system models. However, to ensure that the resulting decisions are trusted, the digital twin must be able to not only predict how a system will respond to a new action or control but also assess the uncertainty associated with that prediction. Much of the literature on optimal decision-making focuses on incorporating uncertainty estimates; notable examples include Markov decision processes and bandit methods. However, these approaches typically reflect one of two extremes: (1) making minimal assumptions about the system and relying entirely on data to estimate uncertainties, or (2) placing strong assumptions on the model and relying on extensive calibration efforts to estimate model parameters a priori. Neither of these two approaches is well suited to incorporating physical models or simulators, and filling this gap is essential to decision-making with digital twins. Interpretability may also be a strong consideration, as first principles–based models may offer decision-makers an understanding of the model parameters and the causal relationships between inputs and outputs.

In data-poor scenarios, the models must necessarily play a greater role within the optimization/control algorithms. There are mature methods for deterministic optimization problems of this nature—for example, in the areas of model predictive control and PDE-constrained optimization. While advances have been made in the stochastic case, solving optimization problems under uncertainty at the scale and model sophistication anticipated for digital twins remains a challenge. A key ingredient for achieving scalability in model-constrained optimization is the availability of sensitivity information (i.e., gradients and possibly higher-order derivatives), often obtained using adjoint methods that scale well for high-dimensional problems. Adjoint methods are powerful, but their implementation is time intensive, requires specialized expertise, and is practically impossible for legacy codebases. Making sensitivity information more readily available would be an enabler for scalable decision-making with model-centric digital twins. This could be achieved by advancing automatic differentiation capabilities, with particular attention to approaches that will be successful for the multiphysics, multiscale, multi-code coupled models that will underlie many digital twins. Variational approaches that compute sensitivity information at the continuous (PDE) level are also emerging as promising tools.

Finding 6-4: Models and data play a synergistic role in digital twin decision-making. The abundance or scarcity of data, complexity of the decision space, need to quantify uncertainty, and need for interpretability are all drivers to be considered in advancing theory and methods for digital twin decision-making.

HUMAN–DIGITAL TWIN INTERACTIONS

Human–computer interaction is the study of the design, evaluation, and implementation of computing systems for human use and the study of the major phenomena surrounding them (Sinha et al. 2010). Research and advances in interfaces and interactions of humans and computers continue to evolve considerably from the earliest Electronic Numerical Integrator and Calculator introduced in 1946 to modern graphical user interfaces (GUIs). However, the nature of human–digital twin interactions poses several unique challenges. The complex and dynamic nature of a digital twin introduces increased challenges around building trust and conveying evolving uncertainty, while also enabling understanding across all individuals who will interact with the digital twin. The contextual details required for digital twins can also introduce challenges in ethics, privacy, ownership, and governance of data around human contributions to and interactions with digital twins.

Use- and User-Centered Design

There is a range of respective roles that humans can play in interactions with digital twins, and the particular role and interaction of the human with a digital twin will influence the design of the digital twin. A key step is defining the intended use of the digital twin and the role of the human with the digital twin and clarifying the required design, development needs, and deployment requirements. The intended use of the digital twin along with the role and responsibilities of the human will help define the necessary data flows (which data at what time interval), range of acceptable uncertainties, and human–computer interaction requirements (e.g., how the data and uncertainties are presented).

In some settings, digital twins will interact with human operators continuously (as opposed to generating an output for subsequent human consumption and action). In these settings, input or feedback from human operators will dynamically alter the state of the digital twin. For example, one might consider a digital twin of a semi-autonomous vehicle that can solicit and incorporate human decisions. In such settings, care is needed to determine how to best solicit human feedback, accounting for human attention fatigue, human insights in new settings as well as human biases, and the most efficacious mechanisms for human feedback (e.g., pairwise or triplet comparisons as opposed to assigning raw numerical scores). Ongoing work in active learning and human–machine co-processing provides many essential insights here. However, as with uncertainty quantification, existing methods do not provide clear mechanisms for incorporating physical or first-principle insights into decision-making processes, and closing this gap is essential for digital twins. To support human–digital twin interactions effectively, focused efforts must be made toward developing implementation science around digital twins, structuring user-centered design of digital twins, and enabling adaptations of human behavior.

Looking to the future, as emerging advances in the field of artificial intelligence (AI) allow for verbal and visual communication of concepts and processes, AI-mediated communications may be incorporated into digital twins to accelerate their creation, maintain their tight alignment with physical twins, and expand their capabilities. Moreover, a decision-making process could leverage a mixed team where AI "team members" manage large amounts of data and resources, provide classifications, and conduct analytical assessments. Human decision-makers could use this information to determine the course of action; in this way, AI components could assist in reducing the cognitive load on the human while enhancing the decision-making capabilities of the team.

Human Interaction with Digital Twin Data, Predictions, and Uncertainty

Effective visualization and communication of digital twin data, assumptions, and uncertainty are critical to ensure that the human user understands the content, context, and limitations that need to be considered in the resulting decisions. While opportunities for data visualization have expanded considerably over recent years, including the integration of GUIs and virtual reality capabilities, the understanding and visualization of the content in context, including the related uncertainties, remains difficult to capture; effective methods for communicating uncertainties necessitate further exploration.

Beyond the objective understanding of the uncertainties around the digital twin predictions, circumstantial and contextual factors including the magnitude of impact as well as the time urgency can influence human perception and decision-making amidst human–digital twin interactions. For instance, the prediction of daily temperature ranges for the week is likely to be received differently from the prediction and related uncertainty in the course of a hurricane or tornado, due to the magnitude of impact of the uncertainty and the immediate time-sensitive decisions that need to be made with this information. Similar parallels can be drawn for patients having their weight or blood pressure progress tracked over time amidst lifestyle or medication interventions, in which case some range of error is likely to be considered acceptable. In contrast, uncertainties in predictions of outcomes of interventions for a diagnosis of cancer would most likely be perceived and considered very differently by individuals due to the gravity of the situation and magnitude of impact of treatment decisions. While there is general recognition that the selected content and context (including uncertainties) around the presentation of information by the digital twin to the user will impact decision-making, there is limited research on the impact of the content, context, and mode of human–digital twin interaction on the resulting decisions. Uncertainty quantification advancements in recent decades provide methods to identify the sources of uncertainty, and in some settings an opportunity the reduce uncertainty (NRC 2012). Understanding uncertainty is not just a technical requirement but a foundational aspect of building trust and reliability among end users. When stakeholders are aware of the levels of uncertainty in the data or model predictions, they can make more nuanced and informed decisions. Additionally, the manner in which uncertainty is communicated can itself be a subject of research and innovation, involving the fields of user experience design, cognitive psychology, and even ethics. Techniques like visualization, confidence intervals, or interactive dashboards can be deployed to make the communication of uncertainty more effective and user-friendly.

Conclusion 6-2: Communicating uncertainty to end users is important for digital twin decision support.

Establishing Trust in Digital Twins

There are many aspects that add to the complexity of establishing trust for digital twins. As with most models, trust in a digital twin need not—and probably should not—be absolute. A digital twin cannot replace reality, but it might provide adequate insight to help a decision-maker. The end user needs to understand the parameter ranges in which the digital twin is reliable and trustworthy as well as what aspects of the digital twin outputs carry what level of trust. For example, a digital twin could be considered trustworthy in its representation of some physical behaviors but not of all of them. The interdisciplinary and interactive nature within the digital twin and across various stakeholders of the digital twin adds an additional layer of complexity to trust. Trauer et al. (2023) present three basic types of stakeholders in the context of a digital twin: the digital twin supplier, a user of the digital twin, and partners of the user. The authors describe how each of these stakeholders requires trust around different aspects related to the digital twin.

Similar to challenges of other artificial intelligence decision-support tools, interpretable methods are essential to establish trust, as humans generally need more than a black-box recommendation from a digital twin. However, an added complexity for digital twins is that as they change over time in response to new data and, in turn, experience changes in the recommended decisions, a way to communicate this evolution to the human user is needed—providing transparency of what has changed. For humans to believe and act on input from digital twins, the insights need to be presented with the acknowledged context of the physical elements including the constraints on the insights so that humans can understand and align the insights with the physical world. For example, a digital twin of electric grids should include the insights in the context of power flow mechanics in the electric grids.

As noted above, a key aspect to this transparency is the presentation of the insights with uncertainty quantification to the human in a manner that can be understood and considered in the decision-making process.

Finding 6-5: In addition to providing outputs that are interpretable, digital twins need to clearly communicate any updates and the corresponding changes to the VVUQ results to the user in order to engender trust.

Human Interactions with Digital Twins for Data Generation and Collection

The technology to collect data across the human–digital interface that could ultimately support human–digital twin interactions is growing rapidly. This includes using the digital twin software or interface to capture human interactions—for example, capturing the number of button clicks and mouse movements to facial expressions and gaze tracking. Data on human behavior such as biometrics being captured across various commercial devices that track step count, heart rate, and beyond can also inform digital twins for various applications to improve health care and wellness. With growing utilization of augmented reality and virtual reality, the collection of human interactions in the digital space will continue to increase and serve as a source of data for human–digital twin interactions. The data gathered within these interactions can also inform what and how future data capture is integrated into the digital twin (e.g., timing of assessments or measurements, introduction of new biosensors for humans interacting with digital twins).

As highlighted in other sections of this report, data acquisition and assimilation are major challenges for digital twins. Semantic interoperability challenges arise as a result of human–digital twin interactions. Moreover, data quality suffers from temporal inconsistencies; changes in data storage, acquisition, and assimilation practices; and the evolution of supporting technology. Despite various efforts across multiple organizations to establish standards, the adoption of standard terminologies and ontologies and the capture of contextual information have been slow even within domain areas. As we look to the multiscale, multidisciplinary interactions required for digital twins, the need to harmonize across domains magnifies the challenge. While the capture of enough contextual detail in the metadata is critical for ensuring appropriate inference and semantic interoperability, the inclusion of increasing details may pose risks to privacy.

ETHICS AND SOCIAL IMPLICATIONS

Human–digital twin interactions raise considerations around ethics across various aspects including privacy, ethical bias propagation, and the influence of human–digital twin interactions on the evolution of human society. A digital twin of a human or a component (e.g., organ) of a human is inherently identifiable, and this poses questions around privacy and ownership as well as rights to access and the ethical responsibility of all who have access to this information. Individuals may be pressured or even coerced to provide the data being collected for the digital twin (Swartz 2018). For instance, in health care, the digital twin may be of a patient, but there are multiple humans-in-the-loop interacting with the digital twin including the patient and perhaps caregiver(s), providers that could encompass a multidisciplinary team, and other support staff who are generating data that feed into the digital twin. The governance around the data of all these interactions from all these humans remains unclear.

When considering human–digital twin interactions in health care, for instance, models may yield discriminatory results that may result from biases in the training data set (Obermeyer et al. 2019). Additionally, the developers may introduce biased views on illness and health into the digital twin that could influence the outputs. For example, a biased view grounded in a victim-blaming culture or an overly focused view on preconceived "health"-related factors that ignores other socioeconomic or environmental factors may clearly limit the ability to follow the recommendations to improve health (Marantz 1990). For example, patients who are not compliant with taking the recommended treatment may be viewed in a negative manner without considering potential financial, geographical, or other social limitations such as money to fill a prescription, time away from work to attend a therapy session, or even availability of fresh fruits and vegetables in their geographical proximity.

Conclusion 6-3: While the capture of enough contextual detail in the metadata is critical for ensuring appropriate inference and interoperability, the inclusion of increasing details may pose emerging privacy and security risks. This aggregation of potentially sensitive and personalized data and models is particularly challenging for digital twins. A digital twin of a human or component of a human is inherently identifiable, and this poses questions around privacy and ownership as well as rights to access.

Conclusion 6-4: Models may yield discriminatory results from biases of the training data sets or introduced biases from those developing the models. The human–digital twin interaction may result in increased or decreased bias in the decisions that are made.

KEY GAPS, NEEDS, AND OPPORTUNITIES

In Table 6-1, the committee highlights key gaps, needs, and opportunities for enabling the feedback flow from the virtual representation to the physical counterpart of a digital twin. This is not meant to be an exhaustive list of all opportunities presented in the chapter. For the purposes of this report, prioritization of a gap is indicated by 1 or 2. While the committee believes all of the gaps listed are of high priority, gaps marked 1 may benefit from initial investment before moving on to gaps marked with a priority of 2.

TABLE 6-1 Key Gaps, Needs, and Opportunities for Enabling the Feedback Flow from the Virtual Representation to the Physical Counterpart of a Digital Twin

Maturity	Priority
Early and Preliminary Stages	
Scalable methods are needed for goal-oriented sensor steering and optimal experimental design that encompass the full sense–assimilate–predict–control–steer cycle while accounting for uncertainty.	1
Trusted machine learning and surrogate models that meet the computational and temporal requirements for digital twin real-time decision-making are needed.	1
Scalable methods to achieve dynamic adaptation in digital twin decision-making are needed.	2
Theory and methods to achieve trusted decisions and quantified uncertainty for data-centric digital twins employing empirical and hybrid models are needed.	1
Methods and tools to make sensitivity information more readily available for model-centric digital twins, including automatic differentiation capabilities that will be successful for multiphysics, multiscale, multi-code digital twin virtual representations, are needed.	1
Research on and development of implementation science around digital twins, user-centered design of digital twins, and adaptations of human behavior that enable effective human–digital twin teaming are needed. Certain domains and sectors have had more success, such as in the Department of Defense.	1
Uncertainty visualization is key to ensuring that uncertainties are appropriately considered in the human–digital twin interaction and resulting decisions, but development of effective approaches for presenting uncertainty remains a gap.	2
While there is general recognition that the selected content and context (including uncertainties) around the presentation of information by the digital twin to the user will impact decision-making, there is limited research on the impact of the content, context, and mode of human–digital twin interaction on the resulting decisions.	1
Some Research Base Exists But Additional Investment Required	
Scalable and efficient optimization and uncertainty quantification methods that handle non-differentiable functions that arise with many risk metrics are lacking.	2
Research Base Exists with Opportunities to Advance Digital Twins	
Methods and processes to incorporate state-of-the-art risk metrics in practical science, engineering, and medicine digital twin decision-making contexts are needed.	2

REFERENCES

Brunke, L., M. Greeff, A.W. Hall, Z. Yuan, S. Zhou, J. Panerati, and A.P. Schoellig. 2022. "Safe Learning in Robotics: From Learning-Based Control to Safe Reinforcement Learning." *Annual Review of Control, Robotics, and Autonomous Systems* 5:411–444.

Ferrari, A. 2023. "Building Robust Digital Twins." Presentation to the Committee on Foundational Research Gaps and Future Directions for Digital Twins. April 24. Washington, DC.

Ghattas, O. 2022. "A Perspective on Foundational Research Gaps and Future Directions for Predictive Digital Twins." Presentation to the Committee on Foundational Research Gaps and Future Directions for Digital Twins. November 15. Washington, DC.

Ghattas, O. 2023. Presentation to the Workshop on Digital Twins in Atmospheric, Climate, and Sustainability Science. February 1. Washington, DC.

Isele, D., A. Nakhaei, and K. Fujimura. 2018. "Safe Reinforcement Learning on Autonomous Vehicles." Pp. 1–6 in *2018 IEEE/RSJ International Conference on Intelligent Robots and Systems (IROS)*. Madrid, Spain.

Marantz, P.R. 1990. "Blaming the Victim: The Negative Consequence of Preventive Medicine." *American Journal of Public Health* 80(10):1186–1187.

NRC (National Research Council). 2012. *Assessing the Reliability of Complex Models: Mathematical and Statistical Foundations of Verification, Validation, and Uncertainty Quantification*. Washington, DC: The National Academies Press.

Obermeyer, Z., B. Powers, C. Vogeli, and S. Mullainathan. 2019. "Dissecting Racial Bias in an Algorithm Used to Manage the Health of Populations." *Science* 366(6464):447–453.

Royset, J.O. 2023. "Risk-Adaptive Decision-making and Learning." Presentation to the Committee on Foundational Research Gaps and Future Directions for Digital Twins. February 13. Washington, DC.

Sinha, G., R. Shahi, and M. Shankar. 2010. "Human Computer Interaction." Pp. 1–4 in *Proceedings of the 2010 3rd International Conference on Emerging Trends in Engineering and Technology (ICETET '10)*. IEEE Computer Society.

Swartz, A.K. 2018. "Smart Pills for Psychosis: The Tricky Ethical Challenges of Digital Medicine for Serious Mental Illness." *The American Journal of Bioethics* 18(9):65–67.

Trauer, J., D.P. Mac, M. Mörtl, and M. Zimmermann. 2023. "A Digital Twin Business Modelling Approach." Pp. 121–130 in *Proceedings of the International Conference on Engineering Design (ICED23)*. Bordeaux: Cambridge University Press.

7

Toward Scalable and Sustainable Digital Twins

Realizing the societal benefits of digital twins in fields such as biomedicine, climate sciences, and engineering will require both incremental and more dramatic research advances in cross-disciplinary approaches and accompanying infrastructure, both technical and human. The development and evolution of digital twins rely on bridging the fundamental research challenges in statistics, mathematics, and computing as described in Chapters 3 through 6. Bringing complex digital twins to fruition necessitates robust and reliable yet agile and adaptable integration of all these disparate pieces.

This chapter discusses crosscutting issues such as evolution and sustainability of the digital twin; translation of digital twin practices between different domains and communities; model and data sharing to advance digital twin methods; and workforce needs and education for digital twin production, maintenance, and use.

EVOLUTION AND SUSTAINABILITY OF A DIGITAL TWIN

As described in Chapter 2, digital twins build on decades of computational, mathematical, statistical, and data science research within and across disciplines as diverse as biology, engineering, physics, and geosciences. The results of this research are encapsulated in core components that form the foundation of a digital twin. Specifically, these components include the virtual representation of a given physical system and bidirectional workflows between the digital twin and the physical counterpart (see Figure 2-1). Response to varying and evolving changes in the physical system, availability of new observational data, updates to the digital model, or changes in the characteristics of the intended use may dictate revision to the workflows of the digital twin. More fundamentally, robust and trustworthy innovation of the digital twin requires that component attributes (e.g., model, data, workflows) are formally described and changeable, and that the integrity and efficacy of the digital twin are preserved across its evolution. The National Science Foundation's (NSF's) Natural Hazards Engineering Research Infrastructure frameworks for data and simulation workflows illustrate an example of a starting point upon which developers could build as they bring trustworthy digital twins to fruition (Zsarnóczay et al. 2023).

Over time, the digital twin will likely need to meet new demands in its use, incorporate new or updated models, and obtain new data from the physical system to maintain its accuracy. Model management is a key consideration to support the digital twin evolution. A digital twin will have its own standards, application programming interfaces, and processes for maintaining and validating bidirectional workflows. It will require disciplined processes to accommodate and validate revisions. Self-monitoring, reporting, tuning, and potentially assisting in its own management are also aspects of digital twin evolution.

In order for a digital twin to reflect temporal and spatial changes in the physical counterpart faithfully, the resulting predictions must be reproducible, incorporate improvements in the virtual representation, and be reusable in scenarios not originally envisioned. This, in turn, requires a design approach to digital twin development and evolution that is holistic, robust, and enduring, yet flexible, composable, and adaptable. Digital twins operate at the convergence of data acquisition (sensors), data generation (models and simulations), large-scale computations (algorithms), visualization, networks, and validation in a secured framework. Digital twins demand the creation of a foundational backbone that, in whole or in part, is reusable across multiple domains (science, engineering, health care, etc.), supports multiple activities (gaining insight, monitoring, decision-making, training, etc.), and serves the needs of multiple roles (analyst, designer, trainee, decision-maker, etc.).

The digital twin benefits from having a well-defined set of services, within a modular, composable, service-oriented architecture, accompanied by robust life-cycle[1] management tools (e.g., revision management). The characteristics of models that can be used within the digital twin workflow should be well specified. The attributes of the bidirectional workflows supported by the digital twin and the attendant resource provisioning required to support timely and reliable decisions are also key considerations for sustaining the digital twin. These workflows, too, may change over time and circumstance, so both formalism and flexibility are required in the design of the digital twin.

Existing literature and documented practices focus on the creation and deployment of digital twins; little attention has been given to sustainability and

---

[1] For the purposes of this report, the committee defines *life cycle* as the "overall process of developing, implementing, and retiring ... systems through a multistep process from initiation, analysis, design, implementation, and maintenance to disposal" as defined in NIST (2009).

maintenance or life-cycle management of digital twins. Communities lack a clear definition of digital twin sustainability and life-cycle management with corresponding needs for maintaining data, software, sensors, and virtual models. These needs may vary across domains.

Conclusion 7-1: The notion of a digital twin has inherent value because it gives an identity to the virtual representation. This makes the virtual representation—the mathematical, statistical, and computational models of the system and its data—an asset that should receive investment and sustainment in ways that parallel investment and sustainment in the physical counterpart.

Recommendation 4: Federal agencies should each conduct an assessment for their major use cases of digital twin needs to maintain and sustain data, software, sensors, and virtual models. These assessments should drive the definition and establishment of new programs similar to the National Science Foundation's Natural Hazards Engineering Research Infrastructure and Cyberinfrastructure for Sustained Scientific Innovation programs. These programs should target specific communities and provide support to sustain, maintain, and manage the life cycle of digital twins beyond their initial creation, recognizing that sustainability is critical to realizing the value of upstream investments in the virtual representations that underlie digital twins.

With respect to data, a key sustainability consideration is the adoption of open, domain-specific and extensible, community data standards for use by the digital twin. These standards should also address both data exchange and curation. Data privacy and additional security considerations may be required within the digital twin depending on the nature of the data and the physical counterpart being represented. While this direction aligns with current trends with respect to research data, it requires additional emphasis across all digital twin domains. Specific findings, gaps, and recommendations for data are addressed in the upcoming section Translation Between Domains.

Scalability of Digital Twin Infrastructure

The successful adoption, deployment, and efficient utilization of digital twins at scale requires a holistic approach to an integrated, scalable, and sustainable hardware and software infrastructure. The holistic system-of-systems characteristic of a digital twin presents the challenge that digital twins must seamlessly operate in a heterogeneous and distributed infrastructure to support a broad spectrum of operational environments, ranging from hand-held mobile devices accessing digital twins "on-the-go" to large-scale centralized high-performance computing (HPC) installations and everything in between. The digital twin may support one or multiple access models such as in situ, nearby, and remote access (Grübel et al. 2022). Digital twins necessitate a move away from fragmented components and toward a trusted and secured single hub of assets that captures, integrates, and delivers disparate bidirectional data flows to produce actionable information. The infrastructure challenge is how to create, access, manage, update, protect, and sustain this hub in today's distributed digital infrastructure.

To support the rapid refresh cycle and persistent interactions required of digital twins, large hardware systems with massive numbers of processors (CPU [central processing unit] and GPU [graphics processing unit]), vast memory, and low-latency, high-bandwidth networks are needed. Newer methods like artificial intelligence (AI) and machine learning (ML) have been able to create new programming approaches that can take full advantage of the newer computational architectures to accelerate their computations. However, not all workloads are good matches for GPU architectures, such as discrete event simulations, due to the generation of non-uniform workloads and resulting potential for performance degradation. Cloud computing may be better suited to handle digital twin components that require dynamically changing computational power, but other components will require a consistent infrastructure.

Simulations for a digital twin will likely require a federation of individual, best-of-breed simulations rather than a single, monolithic simulation software system. A key challenge will be to couple them and orchestrate their integration. To allow a full digital twin ecosystem to develop and thrive, it will be necessary to develop interface definitions and application programming interfaces that enable individual and separate development of simulations that end up running coupled tightly together and influencing each other.

A barrier to realizing digital twins is the speed of the simulations. Classical physical simulations run far slower than real time in order to achieve the desired accuracy. Digital twins will require simulations that can run far faster than real time to simulate what-if scenarios, required fundamental changes in how simulations are designed, deployed, and run. Chapter 3 describes the design and use of surrogate models in order to speed up simulations. To be useful for a deployed digital twin in the field, these simulations may need to be accessible in real time and beyond real time, putting a heavy strain on the communications infrastructure of the end user.

The end user will often be in a position to use classical infrastructure like wired networking and other amenities associated with an office environment to access and control the digital twin. However, a large number of high-value applications for digital twins in areas such as emergency response, natural disaster management and others will depend on mobile infrastructure with significantly lower bandwidth and higher latencies. Mobile networks such 4G and 5G (and even the future 6G) cannot compete with fixed infrastructure, which is where most of the current development happens. Effective access to a digital twin in these constrained conditions will require new methods to present, visualize, and interact with the simulations.

One possible solution to this challenge lies in pushing surrogate models to their limits by deploying them on the end-user devices. This will require truly innovative methods of simplifying or restructuring simulations in order to allow their execution on mobile devices. These kinds of specialized simulations are a good match for ML approaches, which are less concerned with the underlying physical aspects and more with the phenomenological results of the simulations. Surrogate modeling approaches, including ML, are fundamentally asymmetric: developing and training the model is a computationally intensive process that is well suited to large, centralized infrastructure in the form of HPC centers. But the resulting model can, depending on the complexity of the simulation, be fairly compact. This would allow transferring the model into the field quickly, and modern mobile devices already have or, in the near future, are going to have sufficient computational capacity to execute learned models.

In addition to these extreme cases (model running on HPC versus model running on mobile device), intermediate solutions are possible, in which the model is run on an edge system close to the end user, that can provide higher computational capacity than the mobile device, but at a closer network distance and therefore at lower latency. Such capacity could make it feasible to run complex models in time-sensitive or resource-constrained environments.

Developers do not need to replicate infrastructure at all sites where the digital twin needs to be utilized. Instead, a distributed heterogeneous infrastructure capable of routing data and computational resources to all places where the digital twin is used may be preferable. Sustaining a robust, flexible, dynamic, accessible, and secure digital twin is a key consideration for creators, funders, and the diverse community of stakeholders.

CROSSCUTTING DIGITAL TWIN CHALLENGES AND TRANSLATION ACROSS DOMAINS

As can be seen throughout this report, there are domain-specific and even use-specific digital twin challenges, but there are also many elements that cut across domains and use cases. Many common challenges were noted across the three information-gathering workshops on atmospheric, climate, and sustainability sciences (NASEM 2023a); biomedical sciences (NASEM 2023b); and engineering (NASEM 2023c). In particular, the bidirectional interaction between virtual and physical together with the need for on-demand or continual access to the digital twin present a set of challenges that share common foundational research gaps—even if these challenges manifest in different ways in different settings. This blend of domain specificity and commonality can be seen in each of the elements of the digital twin ecosystem. When it comes to the digital twin virtual representation, advancing the models themselves is necessarily domain specific, but advancing the digital twin enablers of hybrid modeling and surrogate modeling embodies shared challenges that crosscut domains. Similarly, for the physical counterpart, many of the challenges around sensor technologies and data are domain specific, but issues around handling and fusing multimodal data, sharing of data, and advancing data curation practices embody shared challenges that crosscut domains. When it comes to the physical-to-virtual and virtual-to-physical flows that are so integral to the notion of a digital twin, there is an opportunity to advance data assimilation, inverse methods, control, and sensor-steering methodologies that are applicable across domains, while at the same time recognizing domain-specific needs, especially as they relate to the domain-specific nature of decision-making. Finally, verification, validation, and uncertainty quantification (VVUQ) is another area that has some domain-specific needs but represents a significant opportunity to advance digital twin VVUQ methods and practices for digital twins in ways that translate across domains.

As stakeholders consider architecting programs that balance these domain-specific needs with cross-domain opportunities, it is important to recognize that different domains have different levels of maturity with respect to the different elements of the digital twin ecosystem. For example, the Earth system science community is a leader in data assimilation (NASEM 2023a); many fields of engineering are leaders in integrating VVUQ into simulation-based decision-making (NASEM 2023c); and the medical community has a strong culture of prioritizing the role of a human decision-maker when advancing new technologies (NASEM 2023b). Cross-domain interaction through the common lens of digital twins is an opportunity to share, learn, and cross-fertilize.

Throughout the committee's information gathering, it noted several nascent digital twin efforts within communities but very few that spanned different domains. The committee also noted that in some areas, large-scale efforts in Europe are being initiated, which offer opportunity both to collaborate and to coordinate efforts (e.g., Destination Earth,[2] European Virtual Human Twin [EDITH 2022]).

Finding 7-1: Cross-disciplinary advances in models, data assimilation workflows, model updates, use-specific workflows that integrate VVUQ, and decision frameworks have evolved within disciplinary communities. However, there has not been a concerted effort to examine formally which aspects of the associated software and workflows (e.g., hybrid modeling, surrogate modeling, VVUQ, data assimilation, inverse methods, control) might cross disciplines.

Conclusion 7-2: As the foundations of digital twins are established, it is the ideal time to examine the architecture, interfaces, bidirectional workflows of the virtual twin with the physical counterpart, and community practices in order to make evolutionary advances that benefit all disciplinary communities.

---

[2] The website for Destination Earth is https://destination-earth.eu, accessed June 13, 2023.

Recommendation 5: Agencies should collaboratively and in a coordinated fashion provide cross-disciplinary workshops and venues to foster identification of those aspects of digital twin research and development that would benefit from a common approach and which specific research topics are shared. Such activities should encompass responsible use of digital twins and should necessarily include international collaborators.

Finding 7-2: Both creation and exploration of the applications of digital twins are occurring simultaneously in government, academia, and industry. While many of the envisioned use cases are dissimilar, there is crossover in both use cases and technical need within and among the three sectors. Moreover, it is both likely and desirable that shared learning and selective use of common approaches will accrue benefits to all.

Recommendation 6: Federal agencies should identify targeted areas relevant to their individual or collective missions where collaboration with industry would advance research and translation. Initial examples might include the following:
- Department of Defense—asset management, incorporating the processes and practices employed in the commercial aviation industry for maintenance analysis.
- Department of Energy—energy infrastructure security and improved (efficient and effective) emergency preparedness.
- National Institutes of Health—in silico drug discovery, clinical trials, preventative health care and behavior modification programs, clinical team coordination, and pandemic emergency preparedness.
- National Science Foundation—Directorate for Technology, Innovation and Partnerships programs.

MODEL AND DATA COLLABORATIONS TO ADVANCE DIGITAL TWIN METHODS

While several major models are used within the international climate research community, there is a history of both sharing and coordination of models. Moreover, there is a history of and consistent commitment to data exchange among this international research community that is beneficial to digital twins. This is not as pervasive, for example, among biomedical researchers, even accounting for data privacy. While data and model exchange might not be feasible in domains where national security or privacy might be compromised, model and data collaborations may lead to advancements in digital twin development.

There is a culture of global data collaboration in the weather and climate modeling community that results from the fact that observations from all over the world are needed to get a complete picture of the coupled Earth system—the components of the Earth system, the atmosphere in particular, do not end at political boundaries. A fundamental underlying thread in weather and climate research is that, knowing the governing equations of atmospheric behavior (e.g., circulation, thermodynamics, transports of constituents), forecasts of the future state of the atmosphere can be made from a complete set of observations of its current state and the underlying surface and energy exchanges with Earth's space environment. Because the atmospheric circulation develops and advects properties of the atmosphere on time scales of days, forecasts for a given locality depend on the state of the system upstream within the recent past; observations made over a neighboring jurisdiction are necessary to make a forecast.

The necessity for data exchange among nations was recognized in the mid-19th century, and a rich network of data observations and high-speed communications has developed over the decades since (Riishojgaard et al. 2021). Today, global observations of the atmosphere and ocean are routinely taken from a variety of instrument platforms—including surface land-based stations, rawinsondes, commercial and research ships, aircraft, and satellites—and instantaneously telemetered to a set of national meteorological and hydrological services that act as hubs for the data via the Global Telecommunication System.[3] A vast infrastructure supports this network. The real-time transmission of data is backed up by a network of archival facilities, such as the National Centers for Environmental Information,[4] which archives and makes publicly available more than 700 TB of Earth observations and analyses each month. The global network of atmospheric observations used for weather forecasting is augmented for longer records in the Global Climate Observing System.[5]

The evolution of the global observing network has been advised by the use of the observations to initialize weather and climate forecasts via data assimilation in which a model of the Earth system (or one of its components) is used to generate an estimate of the current state of the system, which is then optimally combined with the observations to produce an initial condition for a forecast. The combination of the observing system, data assimilation system, and forward model conforms to the definition of a digital twin insofar as data from the real world are constantly being used to update the model, and the performance of the model is constantly being used to refine the observing system.

With respect to coordination of models, the United States Global Change Research Program (USGCRP)[6] is an interagency body of the federal govern-

---

[3] The website for the Global Telecommunication System is https://public.wmo.int/en/programmes/global-telecommunication-system, accessed September 20, 2023.

[4] The website for the National Centers for Environmental Information is https://www.ncei.noaa.gov, accessed June 26, 2023.

[5] The website for the Global Climate Observing System is https://public.wmo.int/en/programmes/global-climate-observing-system, accessed September 20, 2023.

[6] The website for the U.S. Global Change Research Program is https://www.globalchange.gov/about, accessed July 7, 2023.

ment established by Congress in 1990 to coordinate activities relating to global change research, including Earth system modeling. Several federal agencies support independent Earth system modeling.[7] Any of the individual models could be viewed as a digital twin of the Earth's climate system, and at least three of them—Unified Forecast System, Earth System Prediction Capability, and Goddard Earth Observing System—are ingesting observations in real time for prediction purposes. Furthermore, the USGCRP is coordinating these modeling activities through an inter-agency working group (USGCRP n.d.).

While other disciplines have open-source or shared models (e.g., Nanoscale Molecular Dynamics, Gromacs, or Amber within molecular dynamics), few support the breadth in scale and the robust integration of uncertainty quantification that are found in Earth system models and workflows. This lack of coordination greatly inhibits decision support inherent in a digital twin. A greater level of coordination among the multidisciplinary teams of other complex systems, such as biomedical systems, would benefit maturation and adoption of digital twins. More international joint funding mechanisms—such as the Human Frontier Science Program,[8] which aims to solve basic science to solve complex biological problems, and the Human Brain Project in the European Union[9]—offer the size and interdisciplinary makeup to accelerate digital twin development while ensuring cross-system compatibility. The creation of the human genome demonstrates a successful worldwide cooperative effort that advanced common and ambitious research goals. Another objective in establishing these international interdisciplinary collaborations might be to lay the groundwork for establishing norms and standards for evaluation, protection, and sharing of digital twins. While there are similar examples of community data sets (e.g., Modified National Institute of Standards and Technology, National Health and Nutrition Examination Survey, Human Microbiome Project, Laser Interferometer Gravitational-Wave Observatory,[10] U.S. Geological Survey Earthquake Hazards Program[11]), few communities have the comprehensive and shared set of global observational data available to the Earth system modeling community, which obtains a synoptic view of the planet's coupled components. Such a level of sharing requires the

---

[7] Including the Department of Energy (https://e3sm.org), the National Science Foundation (https://www.cesm.ucar.edu), the National Aeronautics and Space Administration (https://www.giss.nasa.gov/tools/modelE), the National Oceanic and Atmospheric Administration (https://www.gfdl.noaa.gov/model-development and https://ufscommunity.org), and the Navy (Barton et al. 2020; https://doi.org/10.1029/2020EA001199).

[8] The website for the International Human Frontier Science Program Organization is https://www.hfsp.org, accessed June 20, 2023.

[9] The website for the Human Brain Project is https://www.humanbrainproject.eu/en, accessed June 26, 2023.

[10] The website for the Laser Interferometer Gravitational-Wave Observatory data set is https://www.ligo.caltech.edu/page/ligo-data, accessed June 14, 2023.

[11] The website for the U.S. Geological Survey Earthquake Hazards Program is https://www.usgs.gov/programs/earthquake-hazards, accessed June 26, 2023.

establishment of agreements among a diverse group of stakeholders that must be reviewed and revised as circumstances evolve. Incentives and frameworks (including frameworks that go beyond mere aggregation of de-identified data) for comprehensive data collaborations, standardization of data and metadata, and model collaborations would likely aid in this effort.

Conclusion 7-3: Open global data and model exchange has led to more rapid advancement of predictive capability within the Earth system sciences. These collaborative efforts benefit both research and operational communities (e.g., global and regional weather forecasting, anticipation and response to extreme weather events).

Conclusion 7-4: Fostering a culture of collaborative exchange of data and models that incorporate context through metadata and provenance in digital twin–relevant disciplines could accelerate progress in the development and application of digital twins.

Recommendation 7: In defining new digital twin research efforts, federal agencies should, in the context of their current and future mission priorities, (1) seed the establishment of forums to facilitate good practices for effective collaborative exchange of data and models across disciplines and domains, while addressing the growing privacy and ethics demands of digital twins; (2) foster and/or require collaborative exchange of data and models; and (3) explicitly consider the role for collaboration and coordination with international bodies.

PREPARING AN INTERDISCIPLINARY WORKFORCE FOR DIGITAL TWINS

While digital twins present opportunity for dramatic improvement in accurate predictions, decision support, and control of highly complex natural and engineered systems, successful adoption of digital twins and their future progress hinge on the appropriate *education* and *training* of the workforce. This includes formalizing, nurturing, and growing critical computational, mathematical, and engineering skill sets at the intersection of disciplines such as biology, chemistry, and physics. These critical skill sets include but are not limited to "systems engineering, systems thinking and architecting, data analytics, ML/AI, statistics/probabilistic, modeling and simulation, uncertainty quantification, computational mathematics, and decision science" (AIAA Digital Engineering Integration Committee 2020). Unfortunately, few academic curricula foster such a broad range of skills.

New interdisciplinary programs encompassing the aforementioned disciplines, ideally informed by the perspectives of industry and federal partners, could help prepare tomorrow's workforce (AIAA Digital Engineering Integration Committee 2020). Successful workforce training programs for digital twins require a multipronged approach, with new interdisciplinary programs within the academic system (e.g., interdisciplinary degree programs or online certificate programs) led by governmental agencies (e.g., fellowships at national laboratories) or offered in collaboration with industrial partners (e.g., internships).

In the context of workforce development for digital twins, the committee identifies three core areas where foundational improvements can have significant impact: interdisciplinary degrees, research training programs, and faculty engagement.

Interdisciplinary Degrees

Progress in both advancing and adopting digital twins requires interdisciplinary education. Workforce development may require new curricula, but it can be difficult to create interdisciplinary curricula within the existing structure of most universities. Crosscutting and interdisciplinary research that is foundational, rather than only applied, requires incentives and specific support for a culture change that can foster horizontal research across institutions.

Workforce needs for digital twins require students who are educated across the boundaries of computing, mathematics and statistics, and domain sciences. This is the domain of the interdisciplinary fields of computational science and engineering (CSE) and, more recently, data science and engineering (DSE). Interdisciplinary educational degree programs in CSE and DSE have been growing in number across the nation but remain less common than traditional disciplinary programs, especially at the undergraduate level. Traditional academic structures, which tend to reward vertical excellence over interdisciplinary achievement, are often not well suited to cultivate interdisciplinary training.

Some good models for overcoming disciplinary silos and barriers at universities include new interdisciplinary majors (e.g., the Computational Modeling and Data Analytics program at Virginia Tech) and new classes. There could also be interdisciplinary centers that could serve as a fulcrum for engagement with other universities, government agencies, and industry partners (e.g., the Interdisciplinary Research Institutes at Georgia Tech). The national laboratories provide a good model for interdisciplinary research. However, a driving force (e.g., an overarching problem) to serve as a catalyst for such initiatives is needed. Programs such as the NSF Artificial Intelligence Institutes support a culture of interdisciplinary research, but additional incentives are needed to foster broader recognition and adoption of interdisciplinary research within academic institutions.

Finding 7-3: Interdisciplinary degrees and curricula that span computational, data, mathematical, and domain sciences are foundational to creating a workforce to advance both development and use of digital twins. This need crosses fundamental and applied research in all sectors: academia, government, and industry.

Recommendation 8: Within the next year, federal agencies should organize workshops with participants from industry and academia to identify barriers, explore potential implementation pathways, and incentivize the creation of interdisciplinary degrees at the bachelor's, master's, and doctoral levels.

Research Training Programs

In order for the necessary growth in both research and standardization of digital twin approaches across domains as diverse as climate science, engineering, and biomedicine to occur, a targeted interdisciplinary effort must be undertaken that engages academia, industry, and government as part of the digital twin community.

Two types of training will be required to advance the benefits of and opportunities for digital twins. One type of training focuses on the exploration and development of new capabilities within digital twins, while the other type of training focuses on effective use of digital twins. This training will be dynamic as digital twins mature and will need to occur at various levels, including at community colleges and trade schools (e.g., certificate programs).

There are few examples of successful research training programs for interdisciplinary work. The Department of Energy's (DOE's) Computational Science Graduate Fellowship program[12] is an exemplary model of an effective fellowship program that can be emulated for graduate training of digital twin developers. This program requires that graduate students follow an approved interdisciplinary course of study that includes graduate courses in scientific or engineering disciplines, computer science, and applied mathematics. The fellows are required to spend a summer at a DOE laboratory conducting research in an area different from their thesis, consistent with the interdisciplinary emphasis of the program. More recently, NSF offers student programs for supporting internships in industry and at national laboratories.

Federal agencies can also stimulate the digital twin interdisciplinary aspects through federally funded research and development centers, institutes, and fundamental research at the intersection of disciplines. These efforts provide stimulation both for small-team efforts as well as at-scale research. One might also nurture cross-sector collaborations among industry, academia, and federal agencies to benefit from the strengths of each. In such collaborations, parties must

---

[12] The website for the Computational Science Graduate Fellowship program is https://science.osti.gov/ascr/CSGF, accessed July 3, 2023.

be cognizant of different expectations and strengths (e.g., reciprocal benefits), different time scales and timelines, and different organizational cultures.

Faculty Engagement

A successful interdisciplinary research, teaching, and training program requires that the faculty perform interdisciplinary research. The resulting expertise is reflected in course content as well as in digital twin research contributions. It is impractical to have CSE faculty cover the breadth and depth of digital twins from foundational methodology to implementation at scale, so instead the committee suggests tutorials, summer schools, and hack-a-thons. Effective digital twin leadership requires disciplinary depth. Our academic institutions nurture this depth well. However, leadership in digital twins also requires the transformative power of interdisciplinary research. It is essential to be able to provide appropriate recognition for interdisciplinary contributions.

Specific to faculty career advancement (i.e., promotion), a challenge is that interdisciplinary research often involves larger teams, with computer science, mathematical, and statistical contributors being in the middle of a long list of authors. As a result, assessing progress and contribution is difficult for both employers (e.g., universities) and funding agencies. Various solutions have been proposed and provide a good starting point to increase the attractiveness of interdisciplinary research (Pohl et al. 2015). The first steps in this direction are done by, for example, the Declaration on Research Assessment,[13] which redefines how scientists should be evaluated for funding and career progress. The European Commission, for example, has recently signed the declaration (Directorate-General for Research and Innovation 2022) and will implement its assessment values for funding awards. Other examples include an increasing number of journals allowing for publication of software and data (e.g., application note by *Bioinformatics* [Oxford Academic n.d.] or *Data*[14]). Github has facilitated the distribution of software tools in a version-controlled manner, allowing the referencing of computer science contributions. Another example is Code Ocean.[15]

Faculty play a critical role in identifying, developing, and implementing interdisciplinary programs; their support and engagement is essential.

KEY GAPS, NEEDS, AND OPPORTUNITIES

In Table 7-1, the committee highlights key gaps, needs, and opportunities for building digital twins that are scalable and sustainable. This is not meant to be an exhaustive list of all opportunities presented in the chapter. For the purposes

---

[13] The website for the Declaration on Research Assessment is https://sfdora.org, accessed September 12, 2023.
[14] The website for *Data* is https://www.mdpi.com/journal/data, accessed July 3, 2023.
[15] The website for Code Ocean is https://codeocean.com, accessed July 2, 2023.

TABLE 7-1 Key Gaps, Needs, and Opportunities for Scalable and Sustainable Digital Twins

Maturity	Priority
Early and Preliminary Stages	
Incentives and frameworks for comprehensive data collaborations, standardization of data and metadata (including across public data sets), and model collaborations are needed. Frameworks are needed that go beyond existing open science frameworks that largely rely on aggregating de-identified data into publicly accessible repositories.	1
Research Ongoing But Limited Results	
Existing literature and documented practices focus on the creation and deployment of digital twins; little attention has been given to sustainability and maintenance or life-cycle management of digital twins. Communities lack a clear definition of digital twin sustainability and life-cycle management with corresponding needs for maintaining data, software, sensors, and virtual models. These needs may vary across domains.	1

of this report, prioritization of a gap is indicated by 1 or 2. While the committee believes all of the below gaps are of high priority, gaps marked 1 may benefit from initial investment before moving on to gaps marked with a priority of 2.

REFERENCES

AIAA (American Institute of Aeronautics and Astronautics) Digital Engineering Integration Committee. 2020. "Digital Twin: Definition & Value." AIAA and AIA Position Paper.
Barton, N., E.J. Metzger, C.A. Reynolds, B. Ruston, C. Rowley, O.M. Smedstad, J.A. Ridout, et al. 2020. "The Navy's Earth System Prediction Capability: A New Global Coupled Atmosphere-Ocean-Sea Ice Prediction System Designed for Daily to Subseasonal Forecasting." *Advancing Earth and Space Sciences* 8(4):1–28.
Directorate-General for Research and Innovation. 2022. "The Commission Signs the Agreement on Reforming Research Assessment and Endorses the San Francisco Declaration on Research Assessment." https://research-and-innovation.ec.europa.eu/news/all-research-and-innovation-news/commission-signs-agreement-reforming-research-assessment-and-endorses-san-francisco-declaration-2022-11-08_en.
EDITH. 2022. "Implementation." https://www.edith-csa.eu/implementation.
Grübel, J., T. Thrash, L. Aguilar, M. Gath-Morad, J. Chatain, R. Sumner, C. Hölscher, and V. Schinazi. 2022. "The Hitchhiker's Guide to Fused Twins: A Review of Access to Digital Twins In Situ in Smart Cities." *Remote Sensing* 14(13):3095.
NASEM (National Academies of Sciences, Engineering, and Medicine). 2023a. *Opportunities and Challenges for Digital Twins in Atmospheric and Climate Sciences: Proceedings of a Workshop—in Brief.* Washington, DC: The National Academies Press.
NASEM. 2023b. *Opportunities and Challenges for Digital Twins in Biomedical Research: Proceedings of a Workshop—in Brief.* Washington, DC: The National Academies Press.
NASEM. 2023c. *Opportunities and Challenges for Digital Twins in Engineering: Proceedings of a Workshop—in Brief.* Washington, DC: The National Academies Press.

NIST (National Institute of Standards and Technology). 2009. "The System Development Life Cycle (SDLC)." *ITL Bulletin*. https://tsapps.nist.gov/publication/get_pdf.cfm?pub_id=902622.

Oxford Academic. n.d. "Instructions to Authors." https://academic.oup.com/bioinformatics/pages/instructions_for_authors#Scope. Accessed June 29, 2023.

Pohl, C., G. Wuelser, P. Bebi, H. Bugmann, A. Buttler, C. Elkin, A. Grêt-Regamey, et al. 2015. "How to Successfully Publish Interdisciplinary Research: Learning from an Ecology and Society Special Feature." *Ecology and Society* 20(2).

Riishojgaard, L.P., J. Zillman, A. Simmons, and J. Eyre. 2021. "WMO Data Exchange—Background, History and Impact." *World Meteorological Organization* 70(2).

USGCRP (U.S. Global Change Research Program). n.d. "Interagency Group on Integrative Modeling." https://www.globalchange.gov/about/iwgs/igim. Accessed June 7, 2023.

Zsarnóczay, A., G.G. Deierlein, C.J. Williams, T.L. Kijewski-Correa, A-M. Esnard, L. Lowes, and L. Johnson. 2023. "Community Perspectives on Simulation and Data Needs for the Study of Natural Hazard Impacts and Recovery." *Natural Hazards Review* 24(1):04022042.

8

Summary of Findings, Conclusions, and Recommendations

The following section summarizes the key messages and aggregates the findings, conclusions, and recommendations outlined in this report. This recap highlights that the report's findings, conclusions, and recommendations address broader systemic, translational, and programmatic topics, in addition to more focused digital twin research needs, gaps, and opportunities.

DEFINITION OF A DIGITAL TWIN

This report proposes the following definition of a digital twin:

> A digital twin is a set of virtual information constructs that mimics the structure, context, and behavior of a natural, engineered, or social system (or system-of-systems); is dynamically updated with data from its physical twin; has a predictive capability; and informs decisions that realize value. The bidirectional interaction between the virtual and the physical is central to the digital twin.

SYSTEMIC, TRANSLATIONAL, AND PROGRAMMATIC FINDINGS, CONCLUSIONS, AND RECOMMENDATIONS

The report emphasizes that a digital twin goes beyond simulation to include tighter integration between models, data, and decisions. Of particular importance is the bidirectional interaction, which comprises automated and human-in-the-loop feedback flows of information between the physical system and its virtual representation.

SUMMARY OF FINDINGS, CONCLUSIONS, AND RECOMMENDATIONS 115

Finding 2-1: A digital twin is more than just simulation and modeling.

Conclusion 2-1: The key elements that comprise a digital twin include (1) modeling and simulation to create a virtual representation of a physical counterpart, and (2) a bidirectional interaction between the virtual and the physical. This bidirectional interaction forms a feedback loop that comprises dynamic data-driven model updating (e.g., sensor fusion, inversion, data assimilation) and optimal decision-making (e.g., control, sensor steering).

The report emphasizes the key role of verification, validation, and uncertainty quantification (VVUQ) as essential tasks for the responsible development, implementation, monitoring, and sustainability of digital twins across all domains.

Conclusion 2-2: Digital twins require VVUQ to be a continual process that must adapt to changes in the physical counterpart, digital twin virtual models, data, and the prediction/decision task at hand. A gap exists between the class of problems that has been considered in traditional modeling and simulation settings and the VVUQ problems that will arise for digital twins.

The report highlights the role that VVUQ has played in fostering confidence and establishing boundaries for the use of predictive simulations in critical decision-making, noting that VVUQ will similarly play a central role in establishing trust and guidelines for use for digital twins across domains.

Conclusion 2-3: Despite the growing use of artificial intelligence, machine learning, and empirical modeling in engineering and scientific applications, there is a lack of standards in reporting VVUQ as well as a lack of consideration of confidence in modeling outputs.

Conclusion 2-4: Methods for ensuring continual VVUQ and monitoring of digital twins are required to establish trust. It is critical that VVUQ be deeply embedded in the design, creation, and deployment of digital twins. In future digital twin research developments, VVUQ should play a core role and tight integration should be emphasized. Particular areas of research need include continual verification, continual validation, VVUQ in extrapolatory conditions, and scalable algorithms for complex multiscale, multiphysics, and multi-code digital twin software efforts. There is a need to establish to what extent VVUQ approaches can be incorporated into automated online operations of digital twins and where new approaches to online VVUQ may be required.

Finding 2-2: The Department of Energy Predictive Science Academic Alliance Program has proven an exemplary model for promoting interdisciplinary research in computational science in U.S. research universities and has profoundly affected university cultures and curricula in computational science in the way that VVUQ is infused with scalable computing, programming paradigms on heterogeneous computer systems, and multiphysics and multi-code integration science.

Finding 2-3: Protecting privacy and determining data ownership and liability in complex, heterogeneous digital twin environments are unresolved challenges that pose critical barriers to the responsible development and scaling of digital twins.

Despite the existence of examples of digital twins providing practical impact and value, the sentiment expressed across multiple committee information-gathering sessions is that the publicity around digital twins and digital twin solutions currently outweighs the evidence base of success. Achieving the promise of digital twins requires an integrated and holistic research agenda that advances digital twin foundations.

Conclusion 2-5: Digital twins have been the subject of widespread interest and enthusiasm; it is challenging to separate what is true from what is merely aspirational, due to a lack of agreement across domains and sectors as well as misinformation. It is important to separate the aspirational from the actual to strengthen the credibility of the research in digital twins and to recognize that serious research questions remain in order to achieve the aspirational.

Conclusion 2-6: Realizing the potential of digital twins requires an integrated research agenda that advances each one of the key digital twin elements and, importantly, a holistic perspective of their interdependencies and interactions. This integrated research agenda includes foundational needs that span multiple domains as well as domain-specific needs.

Recommendation 1: Federal agencies should launch new crosscutting programs, such as those listed below, to advance mathematical, statistical, and computational foundations for digital twins. As these new digital twin–focused efforts are created and launched, federal agencies should identify opportunities for cross-agency interactions and facilitate cross-community collaborations where fruitful. An interagency working group may be helpful to ensure coordination.

- *National Science Foundation (NSF)*. NSF should launch a new program focused on mathematical, statistical, and computational foundations for digital twins that cuts across multiple application domains of science and engineering.
    - The scale and scope of this program should be in line with other multidisciplinary NSF programs (e.g., the NSF Artificial Intelligence Institutes) to highlight the technical challenge being solved as well as the emphasis on theoretical foundations being grounded in practical use cases.
    - Ambitious new programs launched by NSF for digital twins should ensure that sufficient resources are allocated to the solicitation so that the technical advancements are evaluated using real-world use cases and testbeds.
    - NSF should encourage collaborations across industry and academia and develop mechanisms to ensure that small and medium-sized industrial and academic institutions can also compete and be successful leading such initiatives.
    - Ideally, this program should be administered and funded by multiple directorates at NSF, ensuring that from inception to sunset, real-world applications in multiple domains guide the theoretical components of the program.
- *Department of Energy (DOE)*. DOE should draw on its unique computational facilities and large instruments coupled with the breadth of its mission as it considers new crosscutting programs in support of digital twin research and development. It is well positioned and experienced in large, interdisciplinary, multi-institutional mathematical, statistical, and computational programs. Moreover, it has demonstrated the ability to advance common foundational capabilities while also maintaining a focus on specific use-driven requirements (e.g., predictive high-fidelity models for high-consequence decision support). This collective ability should be reflected in a digital twin grand challenge research and development vision for DOE that goes beyond the current investments in large-scale simulation to advance and integrate the other digital twin elements, including the physical/virtual bidirectional interaction and high-consequence decision support. This vision, in turn, should guide DOE's approach in establishing new crosscutting programs in mathematical, statistical, and computational foundations for digital twins.
- *National Institutes of Health (NIH)*. NIH should invest in filling the gaps in digital twin technology in areas that are particularly critical to biomedical sciences and medical systems. These include bioethics, handling of measurement errors and temporal varia-

tions in clinical measurements, capture of adequate metadata to enable effective data harmonization, complexities of clinical decision-making with digital twin interactions, safety of closed-loop systems, privacy, and many others. This could be done via new cross-institute programs and expansion of current programs such as the Interagency Modeling and Analysis Group.

- *Department of Defense (DoD).* DoD's Office of the Under Secretary of Defense for Research and Engineering should advance the application of digital twins as an integral part of the digital engineering performed to support system design, performance analysis, developmental and operational testing, operator and force training, and operational maintenance prediction. DoD should also consider using mechanisms such as the Multidisciplinary University Research Initiative and Defense Acquisition University to support research efforts to develop and mature the tools and techniques for the application of digital twins as part of system digital engineering and model-based system engineering processes.
- *Other federal agencies.* Many federal agencies and organizations beyond those listed above can play important roles in the advancement of digital twin research. For example, the National Oceanic and Atmospheric Administration, the National Institute of Standards and Technology, and the National Aeronautics and Space Administration should be included in the discussion of digital twin research and development, drawing on their unique missions and extensive capabilities in the areas of data assimilation and real-time decision support.

VVUQ is a key element of digital twins that necessitates collaborative and interdisciplinary investment.

Recommendation 2: Federal agencies should ensure that verification, validation, and uncertainty quantification (VVUQ) is an integral part of new digital twin programs. In crafting programs to advance the digital twin VVUQ research agenda, federal agencies should pay attention to the importance of (1) overarching complex multiscale, multiphysics problems as catalysts to promote interdisciplinary cooperation; (2) the availability and effective use of data and computational resources; (3) collaborations between academia and mission-driven government laboratories and agencies; and (4) opportunities to include digital twin VVUQ in educational programs. Federal agencies should consider the Department of Energy Predictive Science Academic Alliance Program as a possible model to emulate.

Existing literature and documented practices focus on the creation and deployment of digital twins; little attention has been given to sustainability and maintenance or life-cycle management of digital twins. Yet sustaining a robust, flexible, dynamic, accessible, and secure digital twin is a key consideration for creators, funders, and the diverse community of stakeholders.

Conclusion 7-1: The notion of a digital twin has inherent value because it gives an identity to the virtual representation. This makes the virtual representation—the mathematical, statistical, and computational models of the system and its data—an asset that should receive investment and sustainment in ways that parallel investment and sustainment in the physical counterpart.

Recommendation 4: Federal agencies should each conduct an assessment for their major use cases of digital twin needs to maintain and sustain data, software, sensors, and virtual models. These assessments should drive the definition and establishment of new programs similar to the National Science Foundation's Natural Hazards Engineering Research Infrastructure and Cyberinfrastructure for Sustained Scientific Innovation programs. These programs should target specific communities and provide support to sustain, maintain, and manage the life cycle of digital twins beyond their initial creation, recognizing that sustainability is critical to realizing the value of upstream investments in the virtual representations that underlie digital twins.

The report calls out a number of domain-specific digital twin challenges, while also noting that there are many research needs and opportunities that cut across domains and use cases. There are significant opportunities to achieve advances in digital twin foundations through translational and collaborative research efforts that bridge domains and sectors.

Finding 7-1: Cross-disciplinary advances in models, data assimilation workflows, model updates, use-specific workflows that integrate VVUQ, and decision frameworks have evolved within disciplinary communities. However, there has not been a concerted effort to examine formally which aspects of the associated software and workflows (e.g., hybrid modeling, surrogate modeling, VVUQ, data assimilation, inverse methods, control) might cross disciplines.

Conclusion 7-2: As the foundations of digital twins are established, it is the ideal time to examine the architecture, interfaces, bidirectional workflows of the virtual twin with the physical counterpart, and community practices in order to make evolutionary advances that benefit all disciplinary communities.

Recommendation 5: Agencies should collaboratively and in a coordinated fashion provide cross-disciplinary workshops and venues to foster identification of those aspects of digital twin research and development that would benefit from a common approach and which specific research topics are shared. Such activities should encompass responsible use of digital twins and should necessarily include international collaborators.

Finding 7-2: Both creation and exploration of the applications of digital twins are occurring simultaneously in government, academia, and industry. While many of the envisioned use cases are dissimilar, there is crossover in both use cases and technical need within and among the three sectors. Moreover, it is both likely and desirable that shared learning and selective use of common approaches will accrue benefits to all.

Recommendation 6: Federal agencies should identify targeted areas relevant to their individual or collective missions where collaboration with industry would advance research and translation. Initial examples might include the following:
- Department of Defense—asset management, incorporating the processes and practices employed in the commercial aviation industry for maintenance analysis.
- Department of Energy—energy infrastructure security and improved (efficient and effective) emergency preparedness.
- National Institutes of Health—in silico drug discovery, clinical trials, preventative health care and behavior modification programs, clinical team coordination, and pandemic emergency preparedness.
- National Science Foundation—Directorate for Technology, Innovation and Partnerships programs.

Conclusion 7-3: Open global data and model exchange has led to more rapid advancement of predictive capability within the Earth system sciences. These collaborative efforts benefit both research and operational communities (e.g., global and regional weather forecasting, anticipation and response to extreme weather events).

Conclusion 7-4: Fostering a culture of collaborative exchange of data and models that incorporate context through metadata and provenance in digital twin–relevant disciplines could accelerate progress in the development and application of digital twins.

Recommendation 7: In defining new digital twin research efforts, federal agencies should, in the context of their current and future mission priorities, (1) seed the establishment of forums to facilitate good practices for effective collaborative exchange of data and models across disciplines and domains, while addressing the growing privacy and ethics demands of digital twins; (2) foster and/or require collaborative exchange of data and models; and (3) explicitly consider the role for collaboration and coordination with international bodies.

The report notes that the successful adoption and progress of digital twins hinge on the appropriate education and training of the workforce, recognizing the particular importance of interdisciplinary degrees and curricula.

Finding 7-3: Interdisciplinary degrees and curricula that span computational, data, mathematical, and domain sciences are foundational to creating a workforce to advance both development and use of digital twins. This need crosses fundamental and applied research in all sectors: academia, government, and industry.

Recommendation 8: Within the next year, federal agencies should organize workshops with participants from industry and academia to identify barriers, explore potential implementation pathways, and incentivize the creation of interdisciplinary degrees at the bachelor's, master's, and doctoral levels.

DIGITAL TWIN RESEARCH FINDINGS, CONCLUSIONS, AND RECOMMENDATION

The study identified foundational research needs and opportunities associated with each of the elements of a digital twin: the virtual representation, the physical counterpart, the physical-to-virtual flowpath, and the virtual-to-physical flowpath. These findings, conclusions, and recommendation cover many technical areas, including multiscale, hybrid, and surrogate modeling; system integration and coupling; data acquisition, integration, and interoperability; inverse problems; data assimilation; optimization under uncertainty; automated decision-making; human-in-the-loop decision-making; and human–digital twin interactions.

Conclusion 3-1: A digital twin should be defined at a level of fidelity and resolution that makes it fit for purpose. Important considerations are the required level of fidelity for prediction of the quantities of interest, the available computational resources, and the acceptable cost. This may lead to the digital twin including high-fidelity, simplified, or surrogate models, as well as a mixture thereof. Furthermore, a digital twin may include the ability to represent and query the virtual models at variable levels of resolution and fidelity depending on the particular task at hand and the available resources (e.g., time, computing, bandwidth, data).

Finding 3-1: Approaches to assess modeling fidelity are mathematically mature for some classes of models, such as partial differential equations that represent one discipline or one component of a complex system; however, theory and methods are less mature for assessing the fidelity of other classes of models (particularly empirical models) and coupled multiphysics, multicomponent systems.

Finding 3-2: Different applications of digital twins drive different requirements for modeling fidelity, data, precision, accuracy, visualization, and time-to-solution, yet many of the potential uses of digital twins are currently intractable to realize with existing computational resources.

Finding 3-3: Often, there is a gap between the scales that can be simulated and actionable scales. It is necessary to identify the intersection of simulated and actionable scales in order to support optimizing decisions. The demarcation between resolved and unresolved scales is often determined by available computing resources, not by a priori scientific considerations.

Recommendation 3: In crafting research programs to advance the foundations and applications of digital twins, federal agencies should create mechanisms to provide digital twin researchers with computational resources, recognizing the large existing gap between simulated and actionable scales and the differing levels of maturity of high-performance computing across communities.

Finding 3-4: Advancing mathematical theory and algorithms in both data-driven and multiscale physics-based modeling to reduce computational needs for digital twins is an important complement to increased computing resources.

Finding 3-5: Hybrid modeling approaches that combine data-driven and mechanistic modeling approaches are a productive path forward for meeting the modeling needs of digital twins, but their effectiveness and practical use are limited by key gaps in theory and methods.

Finding 3-6: Integration of component/subsystem digital twins is a pacing item for the digital twin representation of a complex system, especially if different fidelity models are used in the digital twin representation of its components/subsystems.

Finding 3-7: State-of-the-art literature and practice show advances and successes in surrogate modeling for models that form one discipline or one component of a complex system, but theory and methods for surrogates of coupled multiphysics systems are less mature.

Finding 3-8: Digital twins will typically entail high-dimensional parameter spaces. This poses a significant challenge to state-of-the-art surrogate modeling methods.

Finding 3-9: One of the challenges of creating surrogate models for high-dimensional parameter spaces is the cost of generating sufficient training data. Many papers in the literature fail to properly acknowledge and report the excessively high costs (in terms of data, hardware, time, and energy consumption) of training.

Conclusion 3-2: In order for surrogate modeling methods to be viable and scalable for the complex modeling situations arising in digital twins, the cost of surrogate model training, including the cost of generating the training data, must be analyzed and reported when new methods are proposed.

Finding 4-1: Documenting data quality and the metadata that reflect the data provenance is critical.

Finding 4-2: The absence of standardized quality assurance frameworks makes it difficult to compare and validate results across different organizations and systems. This is important for cybersecurity and information and decision sciences. Integrating data from various sources, including Internet of Things devices, sensors, and historical data, can be challenging due to differences in data format, quality, and structure.

Conclusion 4-1: The lack of adopted standards in data generation hinders the interoperability of data required for digital twins. Fundamental challenges include aggregating uncertainty across different data modalities and scales as well as addressing missing data. Strategies for data sharing and collaboration must address challenges such as data ownership and intellectual property issues while maintaining data security and privacy.

Conclusion 5-1: Data assimilation and model updating play central roles in the physical-to-virtual flow of a digital twin. Data assimilation techniques are needed for data streams from different sources and for numerical models with varying levels of uncertainty. A successful digital twin will require the continuous assessment of models. Traceability of model hierarchies and reproducibility of results are not fully considered in existing data assimilation approaches.

Conclusion 5-2: Data assimilation alone lacks the learning ability needed for a digital twin. The integration of data science with tools for digital twins (including inverse problems and data assimilation) will provide opportunities to extract new insights from data.

Finding 6-1: There is a need for digital twins to support complex trade-offs of risk, performance, cost, and computation time in decision-making.

Conclusion 6-1: There is value in digital twins that can optimally design and steer data collection, with the ultimate goal of supporting better decision-making.

Finding 6-2: In many cases, trusted high-fidelity models will not meet the computational requirements to support digital twin decision-making.

Finding 6-3: Theory and methods are being developed for reinforcement learning and for dynamically adaptive optimization and control algorithms. There is an opportunity to connect these advances more strongly to the development of digital twin methodologies.

Finding 6-4: Models and data play a synergistic role in digital twin decision-making. The abundance or scarcity of data, complexity of the decision space, need to quantify uncertainty, and need for interpretability are all drivers to be considered in advancing theory and methods for digital twin decision-making.

Conclusion 6-2: Communicating uncertainty to end users is important for digital twin decision support.

Finding 6-5: In addition to providing outputs that are interpretable, digital twins need to clearly communicate any updates and the corresponding changes to the VVUQ results to the user in order to engender trust.

Conclusion 6-3: While the capture of enough contextual detail in the metadata is critical for ensuring appropriate inference and interoperability, the inclusion of increasing details may pose emerging privacy and security risks. This aggregation of potentially sensitive and personalized data and models is particularly challenging for digital twins. A digital twin of a human or component of a human is inherently identifiable, and this poses questions around privacy and ownership as well as rights to access.

Conclusion 6-4: Models may yield discriminatory results from biases of the training data sets or introduced biases from those developing the models. The human–digital twin interaction may result in increased or decreased bias in the decisions that are made.

Appendixes

A

Statement of Task

A National Academies of Sciences, Engineering, and Medicine–appointed ad hoc committee will identify needs and opportunities to advance the mathematical, statistical, and computational foundations of digital twins in applications across science, medicine, engineering, and society. In so doing, the committee will address the following questions:

Definitions and use cases:

- How are digital twins defined across communities?
- What example use cases demonstrate the value of digital twins that are currently in deployment or development?

Foundational mathematical, statistical, and computational gaps:

- What foundational gaps and research opportunities exist in achieving robust reliable digital twins at scale?
- How do these foundational gaps or opportunities vary across communities and application domains?
- What are the roles of data-driven learning and computational modeling (including mechanistic modeling) in achieving robust and reliable digital twins at scale? What data are needed to enable this modeling?
- What are the needs for validation, verification, and uncertainty quantification of digital twins, and how do these needs vary across communities?

Best practices for digital twin development and use:

- What best or promising practices for digital twins are emerging within and across application domains?
- What opportunities exist for translation of best practices across domains? What challenges exist for translation of best practices across domains?
- How are difficult issues such as verification, validation, reproducibility, certification, security, ethics, consent, and privacy being addressed within domains? What lessons can be applied to other domains where digital twins are nascent?

Moving forward:

- What use cases could advance awareness of and confidence in digital twins?
- What are the key challenges and opportunities in the research, development, and application of advancements in digital twin development and application?
- What roles could stakeholders (e.g., federal research funders, industry, academia, professional societies) play in advancing the development of rigorous scalable foundations of digital twins across scientific, medical, engineering, and societal domains and in encouraging collaboration across communities?

The ad hoc committee will conduct three public workshops and other data-gathering activities to inform its findings, conclusions, and recommendations, which will be provided in the form of a consensus report.

The public workshops will present and discuss the opportunities (e.g., methods, practices, use cases) and challenges for the development and use of digital twins in three separate contexts: biomedical domains, Earth and environmental systems, and engineering. These workshops will bring together diverse stakeholders and experts to address the following topics:

- Definitions and taxonomy of digital twins within the specified domain, including identification of exemplar use cases of digital twins;
- Current methods and promising practices for digital twin development and use at various levels of complexity;
- Key technical challenges and opportunities in the near and long term for digital twin development and use; and
- Opportunities for translation of promising practices from other fields and domains.

The presentations and discussions during the workshops will be summarized and published in three separate Proceedings of a Workshop—in Brief documents.

B

Workshop Agendas

OPPORTUNITIES AND CHALLENGES FOR DIGITAL TWINS IN BIOMEDICAL SCIENCES

January 30, 2023
Virtual[1]

10:00 AM **Welcome and Housekeeping**
Rebecca Willett (The University of Chicago)
Michelle Schwalbe (National Academies) delivering sponsor remarks on behalf of the Department of Energy (DOE), Department of Defense (DoD), National Institutes of Health (NIH), and National Science Foundation (NSF)

10:20 AM **Opening Plenary**
Moderator: Irene Qualters (Los Alamos National Laboratory)
Reinhard Laubenbacher (University of Florida)
Karissa Sanbonmatsu (Los Alamos National Laboratory)

11:20 AM **Break**

11:30 AM **Panel 1: Digital Twins at the Cellular and Molecular Scale**
Moderator: Ines Thiele (National University of Ireland, Galway)
Jeffrey R. Sachs (Merck & Co., Inc.)
Mikael Benson (Karolinska Institute)

---

[1] All time in ET.

Juan Perilla (University of Delaware)
Rommie Amaro (University of California, San Diego)

12:15 PM **Lunch Break**

12:45 PM **Panel 2: Digital Twins at the Organ, Tumor, and Microenvironment Scale**
Moderator: Caroline Chung (MD Anderson Cancer Center)
Tom Yankeelov (The University of Texas at Austin)
Jayashree Kalpathy-Cramer (University of Colorado Denver)
James A. Glazier (Indiana University)
Petros Koumoutsakos (Harvard University)

1:30 PM **Panel 3: Digital Twins at the Whole Human, Multisystem, and Population Scale**
Moderator: Ines Thiele (National University of Ireland, Galway)
Aldo Badano (Food and Drug Administration)
David Miller (Unlearn.AI)
Todd Coleman (Stanford University)
Heiko Enderling (Lee Moffitt Cancer Center)

2:15 PM **Break**

2:30 PM **Panel 4: Connecting Across Scales**
Moderator: Rebecca Willett (The University of Chicago)
Bissan Al-Lazikani (MD Anderson Cancer Center)
Liesbet Geris (University of Liège)
Gary An (University of Vermont)

3:15 PM **Panel 5: Special Session on Privacy, Ethics, and Data Issues**
Moderator: Irene Qualters (Los Alamos National Laboratory)
Jodyn Platt (University of Michigan)
Nathan Price (Thorne HealthTech)
Lara Mangravite (HI-Bio)

3:45 PM **Summary and Convergence**
Caroline Chung (MD Anderson Cancer Center)

DIGITAL TWINS IN ATMOSPHERIC, CLIMATE, AND SUSTAINABILITY SCIENCES

February 1, 2023
Virtual

Session 1—Open

10:00 AM **Welcome and Housekeeping**
Ruby Leung (Pacific Northwest National Laboratory)

10:20 AM **Opening Plenary**
Moderator: Ruby Leung (Pacific Northwest National Laboratory)
Umberto Modigliani (ECMWF)
Venkatramani Balaji (Schmidt Futures)
Mark Taylor (Sandia National Laboratories)

11:20 AM **Break**

11:30 AM **Panel 1: Current Methods and Practices**
Moderators: Colin Parris (GE Digital) and Xinyue Ye (Texas A&M University)
Yuyu Zhou (Iowa State University)
Jean Francois Lamarque (McKinsey)
Gavin A. Schmidt (NASA Goddard Institute for Space Studies)
Christiane Jablonowski (University of Michigan)

12:30 PM **Break**

1:30 PM **Panel 2: Key Technical Challenges and Opportunities**
Moderators: Julianne Chung (Emory University) and Jim Kinter (George Mason University)
Tapio Schneider (California Institute of Technology)
Mike Pritchard (NVIDIA/University of California, Irvine)
Omar Ghattas (The University of Texas at Austin)
Abhinav Saxena (GE Research)
Elizabeth A. Barnes (Colorado State University)

February 2, 2023
Virtual

Session 2—Open

10:00 AM   **Welcome Back and Housekeeping**
             Colin Parris (GE Digital)

10:05 AM   **Panel 3: Translation of Promising Practices to Other Fields**
             Moderators: Julianne Chung (Emory University) and Ruby Leung (Pacific Northwest National Laboratory)
             Cecilia Bitz (University of Washington)
             Anima Anandkumar (California Institute of Technology)
             Emanuele Di Lorenzo (Brown University)
             Anna Michalak (Carnegie Institution for Science)
             John Harlim (The Pennsylvania State University)

11:00 AM   **Break**

11:30 AM   **Panel 4: Special Session on Transparency, Societal Benefit, and Equity**
             Moderators: Colin Parris (GE Digital) and Xinyue Ye (Texas A&M University)
             Amy McGovern (University of Oklahoma)
             Mike Goodchild (University of California, Santa Barbara)
             Mark Asch (Université de Picardie Jules Verne)

12:30 PM   **Summary and Convergence**
             National Academies Study Committee

OPPORTUNITIES AND CHALLENGES FOR DIGITAL TWINS IN ENGINEERING

February 7, 2023
Virtual

10:00 AM   **Welcome and Introduction**
             Conrad Grant (Johns Hopkins University Applied Physics Laboratory)
             Michelle Schwalbe (National Academies) delivering sponsor remarks on behalf of DOE, DoD, NIH, and NSF
             Beth Cady (National Academies)

10:15 AM	**Opening Plenary** Moderator: Derek Bingham (Simon Fraser University) Charles Farrar (Los Alamos National Laboratory)

Over the rest of the workshop day, participants will hear from five use-case speakers as they each address three topics in three moderated panels.

> Elizabeth Baron (Unity Technologies, formerly Ford)
> Karthik Duraisamy (University of Michigan)
> Michael Grieves (Digital Twin Institute)
> S. Michael Gahn (Rolls-Royce)
> Dinakar Deshmukh (General Electric)

10:45 AM	**Panel 1: Current Methods and Practices** Moderator: Parviz Moin (Stanford University)
11:45 AM	**Break**
12:00 PM	**Panel 2: Key Technical Challenges and Opportunities** Moderator: Carolina Cruz-Neira (University of Central Florida)
1:00 PM	**Panel 3: Digital Twin Research and Development Needs and Investment** Moderator: Conrad Tucker (Carnegie Mellon University)
2:00 PM	**Wrap-Up Comments and Adjourn for Day** Conrad Grant (Johns Hopkins University Applied Physics Laboratory)

February 9, 2023
Virtual

10:00 AM	**Welcome Back** Conrad Grant (Johns Hopkins University Applied Physics Laboratory) Michelle Schwalbe (National Academies) delivering sponsor remarks on behalf of DOE, DoD, NIH, and NSF Tho Nguyen (National Academies)
10:15 AM	**Opening Plenary** Moderator: Parviz Moin (Stanford University) Grace Bochenek (University of Central Florida)

Over the rest of the workshop day, participants will hear from four use-case speakers as they each address three topics in three moderated panels.

                José R. Celaya (Schlumberger)
                Pamela Kobryn (Department of Defense)
                Devin Francom (Los Alamos National Laboratory)
                Devin Harris (University of Virginia)

10:45 AM   **Panel 1: Current Methods and Practices**
                    Moderator: Carolina Cruz-Neira (University of Central Florida)

11:45 AM   **Break**

12:00 PM   **Panel 2: Key Technical Challenges and Opportunities**
                    Moderator: Conrad Tucker (Carnegie Mellon University)

1:00 PM   **Panel 3: Digital Twin Research and Development Needs and Investment**
                    Moderator: Derek Bingham (Simon Fraser University)

2:00 PM   **Wrap-Up Comments and Adjourn**
                    Conrad Grant (Johns Hopkins University Applied Physics Laboratory)

C

Opportunities and Challenges for Digital Twins in Atmospheric and Climate Sciences: Proceedings of a Workshop—in Brief

Opportunities and Challenges for Digital Twins in Atmospheric and Climate Sciences: Proceedings of a Workshop—in Brief (National Academies of Sciences, Engineering, and Medicine, The National Academies Press, Washington, DC, 2023) is reprinted here in its entirety. The original Proceedings of a Workshop—in Brief is available at https://doi.org/10.17226/26921.

Opportunities and Challenges for Digital Twins in Atmospheric and Climate Sciences

Proceedings of a Workshop—in Brief

The digital twin is an emerging technology that builds on the convergence of computer science, mathematics, engineering, and the life sciences. Digital twins have the potential to revolutionize atmospheric and climate sciences in particular, as they could be used, for example, to create global-scale interactive models of Earth to predict future weather and climate conditions over longer timescales.

On February 1-2, 2023, the National Academies of Sciences, Engineering, and Medicine hosted a public, virtual workshop to discuss characterizations of digital twins within the context of atmospheric, climate, and sustainability sciences and to identify methods for their development and use. Workshop panelists presented varied definitions and taxonomies of digital twins and highlighted key challenges as well as opportunities to translate promising practices to other fields. The second in a three-part series, this evidence-gathering workshop will inform a National Academies consensus study on research gaps and future directions to advance the mathematical, statistical, and computational foundations of digital twins in applications across science, medicine, engineering, and society.[1]

PLENARY SESSION: DEFINITIONS OF AND VISIONS FOR THE DIGITAL TWIN

During the plenary session, workshop participants heard presentations on the challenges and opportunities for Earth system digital twins, the history of climate modeling and paths toward traceable model hierarchies, and the use of exascale systems for atmospheric digital twins.

Umberto Modigliani, European Centre for Medium-Range Weather Forecasts (ECMWF), the plenary session's first speaker, provided an overview of the European Union's Destination Earth (DestinE) initiative,[2] which aims to create higher-resolution simulations of the Earth system that are based on models that are more realistic than those in the past; better ways to combine observed and simulated information from the Earth system; and interactive and configurable access to data, models, and workflows. More realistic simulations at the global scale could translate to information at the regional scale that better supports decision-making for climate adaptation and mitigation through tight integration and interaction with impact sector models. Now in the first phase (2021-2024) of its 7- to 10-year program, DestinE is beginning

---

[1] To learn more about the study and to watch videos of the workshop presentations, see https://www.nationalacademies.org/our-work/foundational-research-gaps-and-future-directions-for-digital-twins, accessed February 10, 2023.

[2] The website for DestinE is https://digital-strategy.ec.europa.eu/en/policies/destination-earth, accessed March 9, 2023.

to coordinate with several other European initiatives, including Copernicus[3] and the European Open Science Cloud.[4]

Modigliani explained that Earth system digital twins require unprecedented simulation capabilities--for example, ECMWF aspires to have a simulation on the order of 1-4 km at the global scale, which could enable the modeling of small scales of sea ice transformation and the development of more accurate forecasts. Earth system digital twins also demand increased access to observation capabilities, and efforts are under way to develop computing resources to leverage and integrate robust satellite information and other impact sector data. Furthermore, Earth system digital twins require exceptional digital technologies to address the opportunities and challenges associated with extreme-scale computing and big data. DestinE will have access to several pre-exascale systems via EuroHPC, although he pointed out that none of the available computing facilities are solely dedicated to climate change, climate extremes, or geophysical applications.

Modigliani indicated that improved data handling is also critical for the success of Earth system digital twins. The envisioned DestinE simulations could produce 1 PB of data per day, and he said that these data must be accessible to all DestinE users. While ECMWF will handle the modeling and the digital engine infrastructure for the digital twins, the European Organisation for the Exploitation of Meteorological Satellites will manage a data bridge to administer data access via a sophisticated application programming interface (for the computational work) and a data lake, which serves as a repository to store and process data that may be unstructured or structured. Policy makers could eventually access a platform operated by the European Space Agency to better understand how future events (e.g., heat waves) might affect the gross domestic product and to run adaptation scenarios. Current challenges include federating resource management across DestinE and existing infrastructures as well as collaborating across science, technology, and service programs. To be most successful, he emphasized that DestinE would benefit from international partnerships.

Venkatramani Balaji, Schmidt Futures, the plenary session's second speaker, presented a brief overview of the history of climate modeling, starting with a one-dimensional model response to carbon dioxide doubling in 1967. This early work revealed that studying climate change requires conducting simulations over long periods of time. He explained that as the general circulation model evolved in subsequent decades, concerns arose that its columns (where processes that cannot be resolved, such as clouds, reside) place restrictions on model structures, leading to slow progress. Parameterizing clouds is difficult because of their variety and interdependence. Clouds are also sensitive to small-scale dynamics and details of microphysics; parametrizing turbulence is necessary to understand how the small scale interacts with the large scale. He asserted that no resolution exists at which all features can be resolved unless extreme scales of computational capability are reached.

Balaji next discussed how model resolution has evolved over time, describing an example with only ~10 times improvement in 50 years. He suggested that climate models should be capable of 100 simulations of at least 100 years each in 100 days. Uncertainty—including chaotic (i.e., internal variability), scenario (i.e., dependent on human and policy actions), and structural (i.e., imperfect understanding of a system)—is also a key consideration for model design. Furthermore, he stressed that strong scaling is not possible with today's computers, which have become bigger but not faster.

Balaji noted that while various machine learning (ML) methods have been applied successfully for stationary problems (e.g., short forecasts), boundary conditions change over time in climate studies. One cannot use data for training into the future, because no observations of the future exist, he continued, but one could use short-running, high-resolution models to train simpler models to address the non-stationarity of climate. He observed that although digital twins were first introduced for

---

[3] The website for Copernicus is https://www.copernicus.eu/en, accessed April 3, 2023.
[4] The website for the European Open Science Cloud is https://eosc-portal.eu, accessed April 3, 2023.

well-understood engineered systems such as aircraft engines, digital twins are now discussed in the context of imperfectly understood systems. Even if climate and weather diverge further as model-free methods become successful in weather forecasting, he anticipated that theory will still play an important role. Model calibration is needed at any resolution, he added, as are methods for fast sampling of uncertainty. In closing, Balaji remarked that because high-resolution models should not play a dominant role, a hierarchy of models is needed. He suggested that ML offers systematic methods for calibration and emulation in a formal mathematical way to achieve traceable model hierarchies.

Mark Taylor, Sandia National Laboratories, the plenary session's final speaker, discussed the need for more credible cloud physics to achieve digital twins for Earth's atmosphere. For example, a global cloud-resolving model (GCRM) aims to simulate as much as possible using first principles. Although GCRMs are the "backbone" of any digital twin system, they require exascale resources to obtain necessary throughput and confront challenges in ingesting and processing results. He pointed out that computers are becoming more powerful but not necessarily more efficient; although GCRMs are expected to run well on exascale systems, they are expensive, with 1 megawatt-hour per simulated day required.

Taylor described his experience porting the following two types of cloud-resolving models to the Department of Energy's (DOE's) upcoming exascale computers:[5] (1) SCREAM (Simple Cloud-Resolving Energy Exascale Earth System Model [E3SM][6] Atmosphere Model), the GCRM, runs at 3 km, with a goal to run at 1 km; and (2) E3SM-Multiscale Modeling Framework (MMF) is a low-resolution climate model with cloud-resolving "superparameterization."

Taylor explained that SCREAM and E3SM-MMF are running well on graphics processing units (GPUs) systems. However, both use significant resources. GPU nodes consume 4–8 times more power than central processing unit (CPU) nodes but need to be 4–8 times faster. Furthermore, GPUs had only 5 times the performance on a per-watt basis in 10 years compared to CPUs. Until more powerful machines are developed, he anticipated that GCRMs will continue to run on GPUs, because GPUs are 1.5–3 times more efficient than CPUs on a per-watt basis.

Discussion

Incorporating questions from workshop participants, Ruby Leung, Pacific Northwest National Laboratory, moderated a discussion among the plenary speakers. She posed a question about what distinguishes a digital twin from a simulation. Taylor replied that a digital twin is a simulation with more efficient approaches. Balaji posited that digital twins represent the modeling and simulation efforts under way for decades. Modigliani focused on the potential for digital twins to explore what-if scenarios and emphasized that communities seek the approach with the most realistic results. Balaji stressed that in addition to being used for forecasting, digital twins should be employed for experimentation to learn more about climate physics in general.

Leung asked about the statistical considerations and learning paradigms required to combine simulations and observations. Modigliani said that ECMWF uses four-dimensional variational data assimilation to study extreme weather, which leverages both the observations and the model to create an initial state for forecast simulations. Balaji advised using a model that imposes physical consistency on the observations to make sense of disparate observational streams. He added that because "model-free" methods are still trained on model output, more work remains to move directly to observational streams.

Leung inquired about the extent to which GPUs are used to perform physics-based atmospheric simulations. Taylor remarked that E3SM-MMF runs the full atmosphere simulation on GPUs. However, he said that modeling centers do not yet run the full coupled system on GPUs—DOE is ~3 years from achieving that goal with E3SM-MMF. Modigliani emphasized the need to find the

---
[5] Frontier is running now in acceptance testing, and Aurora is expected to run in 2023.
[6] The website for E3SM is https://e3sm.org, accessed March 9, 2023.

APPENDIX C

right version that performs better than CPUs; ECMWF aims to have such a version in a few years.

Leung wondered if a regional-scale model could function as a digital twin that bypasses exascale computing requirements. Modigliani noted that global-scale models are required for any weather forecasting of more than a few days into the future. Leung asked how global digital twin outputs could be interpolated to benefit local decision-making, and Modigliani replied that integrating more specific models into global-scale models guides local decisions—for example, DestinE's extreme weather digital twin includes a global-scale model and a local-level simulation.

PANEL 1: CURRENT METHODS AND PRACTICES
During the first panel, workshop participants heard brief presentations on methods and practices for the use of digital twins.

Yuyu Zhou, Iowa State University, discussed efforts to develop a model to reduce energy use for sustainable urban planning. The current model estimates energy use at the single building level and for an entire city of buildings by integrating the building prototypes, the assessor's parcel data, building footprint data, and building floor numbers, and it includes city-scale auto-calibration to improve the energy use estimate. The model enables the estimation of both the spatial pattern of energy use for each building and the hourly temporal pattern (e.g., the commercial area uses the least energy at midnight and the most at noon). He noted that the model could also be used to investigate the impacts of extreme events (e.g., heat waves) or human activities (e.g., building occupant behavior) on building energy use.

Zhou explained that real-time data from sensors and Internet of Things[7] devices could be assimilated into the building energy use model to develop a digital twin with a more comprehensive, dynamic, and interactive visual representation of a city's buildings. Transportation data and social media data could also be integrated to enhance the representation of building occupant behavior. He

[7] The Internet of Things is the "networking capability that allows information to be sent to and received from objects and devices using the Internet" (*Merriam-Webster*).

asserted that a future digital twin could monitor real-time building energy use and optimize a city's energy performance in real time.

Christiane Jablonowski, University of Michigan, observed that the modeling community has created digital representations of reality for years; however, digital twins could integrate new capabilities such as high-resolution representations of climate system processes as well as advancements in ML and artificial intelligence (AI). She provided a brief overview of the current state of modeling and possible future trajectories, beginning with a description of typical phenomena at temporal and spatial scales (Figure 1). She said that the microscale will never be captured by atmospheric models in any detail, and the current generation of weather models (1–3 km grid scale) has reached its limit at the mesoscale. Modelers are currently most comfortable in the synoptic regime, the site of daily weather. Climate system timescales introduce new uncertainties, as they are in the boundary value problem category, not the initial value problem category. She added that the complexity and resolution of climate models has advanced greatly over the past several decades and will continue to increase with hybrid approaches and ML.

Jablonowski emphasized that selecting the appropriate spatial and temporal scales for digital twins is critical to determine the following: the phenomena that can be represented in a model; the correct equation set for the fluid flow; the required physical parameterizations, conservation principles, and exchange processes (i.e., a good weather model today is not a good climate model tomorrow); the model complexity (e.g., ocean, ice, and chemistry components are often not well tuned); decisions about coupling and related timescales that are key to making trustworthy predictions; and whether AI and ML methods as well as observations could inform and speed up current models.

Jean-François Lamarque, consultant and formerly of the National Center for Atmospheric Research's Climate and Global Dynamics Laboratory, shared the definition of a digital twin, as used in the digital twin Wikipedia entry: "a high-fidelity model of the system, which can

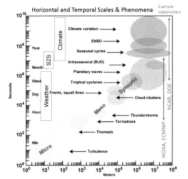 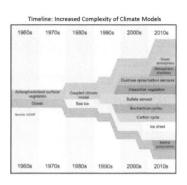

FIGURE 1 Hierarchy of temporal and spatial scales (left) and timeline of model complexities (right). SOURCE: Christiane Jablonowski, University of Michigan, presentation to the workshop. Left: Modified from The COMET Program (the source of this material is the COMET® website at http://meted.ucar.edu of the University Corporation for Atmospheric Research, sponsored in part through cooperative agreements with the National Oceanic and Atmospheric Administration and the Department of Commerce; © 1997–2023 University Corporation for Atmospheric Research; all rights reserved). Right: Copyright University Corporation for Atmospheric Research, licensed under CC BY-NC 4.0 License, via OpenSky.

be used to emulate the actual system … the digital twin concept consists of three distinct parts: the physical object or process and its physical environment, the digital representation of the object or process, and the communication channel between the physical and virtual representations."[8] He emphasized that a research question should determine the tool selection; in the current generation and in the foreseeable future, no single tool exists to answer all relevant questions. Thus, he suggested evaluating the advantages and disadvantages of each tool's approach as well as the timescales of interest. Three current tools include a coarse-resolution Earth system model, a high-resolution climate and weather model, and emulators and ML models. Key questions to evaluate the usefulness of each tool include the accuracy of the digital twin in representing the Earth system, the strength of the communication channel between the physical and digital representations, and the usefulness of the digital twin for climate research.

Gavin A. Schmidt, Goddard Institute for Space Studies of the National Aeronautics and Space Administration, explained that digital twins extend beyond the Earth system modeling that has occurred for decades. digital twins leverage previously untapped data streams,

although whether such streams exist for climate change remains to be seen. Digital twins should explore a full range of possible scenarios, he continued, but only a small number of scenarios can be run with the current technology. A more efficient means to tap into information for downstream uses (e.g., urban planning) would be beneficial, and processes to update information regularly are needed. He added that higher resolution does not necessarily lead to more accurate climate predictions. Furthermore, ML does not overcome systematic biases or reveal missing processes.

Schmidt highlighted the value of improving the skill and usability of climate projections but noted that improvements in initial value problem skill do not automatically enhance boundary value problem skill. He stressed that no "digital twin for climate" exists, but digital twin technology could be used to strengthen climate models; for example, systematic biases could be reduced via ML-driven calibration, new physics-constrained parametrizations could be leveraged, and data usability could be improved.

Discussion

Incorporating questions from workshop participants, Xinyue Ye, Texas A&M University, and Colin Parris, General

---

[8] Wikipedia, "Digital Twin," https://en.wikipedia.org/wiki/Digital_twin.

APPENDIX C

Electric, moderated a discussion among the four panelists. Parris posed a question about how to ensure the quality of both data from varied sources and the output that supports decision-making. Zhou explained that the output from the building energy use model was validated using several sources (e.g., utility company data and survey data). When a digital twin of city buildings emerges in the future, he said that more data could be used to improve the model, and interactions between the physical and the virtual aspects of the digital twin could be enhanced to increase confidence in the data and the output. Schmidt emphasized that data or models are not perfect and uncertainty models help evaluate their effectiveness in representing the real world. Lamarque urged researchers to evaluate multiple observations together, instead of independently, to better understand a system.

Ye inquired about strategies for working with decision-makers to define the requirements of digital twins to ensure that they will be useful. Schmidt remarked that policy makers want to know how the tools could help answer their specific questions but do not need all of the details about how they work. Lamarque mentioned that if too many details are hidden and decision-makers cannot understand the provenance of the information and the uncertainties, a lack of confidence in the results and thus in the decision-making could emerge. Jablonowski asserted that talking to stakeholders about their understanding and helping them learn to interpret uncertainty is critical. Zhou agreed that communication with and education for decision-makers is key.

Parris asked if a full range of scenarios has to be run to make projections for the next decade. Schmidt observed that although some things will not change much in the next 10 years, confidence is lacking in the accuracy of boundary force problems at that timescale. At the 20- to 30-year timescale, the range of scenarios separates, and updating with real-time changes in emissions, technology, and economics is important. He asserted that more bespoke, policy-specific scenarios (e.g., effects of an electric vehicle credit) are needed. Lamarque emphasized the need to find the interplay between the tools that are available and the questions that need to be answered.

Ye posed a question about the ability of the current technology to capture extreme behavior and reliable uncertainty analysis. Schmidt described success in capturing extremes for weather forecasting (e.g., hurricane tracks) but noted that climate change predictions are difficult to validate. Almost all extreme events are at the tail of the distribution; thus, he pointed out that the observational and conceptual challenges of assessing extremes still exist with higher-resolution digital twins.

Parris wondered how the computational challenges for sustainability science compare to those of climate science. Zhou explained that the challenges are similar; for example, many more simulations should be run when scaling from the building to the city level. Cities have thousands of buildings, each with varying conditions (e.g., evapotranspiration and microclimate) that impact energy use uniquely. To develop a more realistic digital twin for a city's buildings, he stressed that improved computation (e.g., edge computing) is essential. Schmidt mentioned that different fields have different perspectives of a "heavy computational load": economists might run a simulation in minutes, while climate scientists might need months or years. Computational capacity is often not used as effectively as possible, he continued, but increased access to high-performance computing could address this issue.

Ye asked about the role of computing capacity problems in the development of digital twins. Schmidt noted that a spread of models with different capabilities and different processes to estimate structural uncertainty and to sample a range of uncertainties will continue to be important. Jablonowski encouraged investments that improve access to extreme-scale computing resources and broaden community engagement in the digital twin endeavor.

PANEL 2: KEY TECHNICAL CHALLENGES AND OPPORTUNITIES
During the second panel, workshop participants heard brief presentations on challenges and opportunities for the use of digital twins.

Tapio Schneider, California Institute of Technology, provided an engineering-specific definition of a digital twin: a digital reproduction of a system that is accurate in detail and is updated in real time with data from the system, allowing for rapid experimentation and prototyping as well as experimental verification and validation where necessary. However, he pointed out that Earth system models are very different: achieving an accurately detailed digital reproduction of the Earth system is not feasible for the purposes of climate prediction, and the use of data for climate prediction (i.e., improving the representation of uncertain processes) is fundamentally different from the use of data for weather prediction (i.e., state estimation)—so continuously updating with data from the real system is less relevant for climate prediction. He suggested a three-pronged approach to advance climate modeling: (1) theory to promote parametric sparsity and generalizability out of observational data distributions, (2) computing to achieve the highest feasible resolution and to generate training data, and (3) calibration and uncertainty quantification by learning from computationally generated and observational data. Learning about processes from diverse data that do not come in the form of input-output pairs of uncertain processes is a key challenge, but he advised that new algorithms that accelerate data assimilation with ML emulators could address this problem.

Mike Pritchard, NVIDIA/University of California, Irvine, offered another description of a digital twin, which he suggested is a surrogate for a deterministic weather prediction system that, once trained, allows for much larger ensembles. Challenges exist in understanding the credibility of these tools as surrogates for the systems on which they are trained, and tests to ensure accountability would be beneficial. He noted that digital twins could help overcome the latency and compression barrier that prevents stakeholders from exploiting the full detail of the Coupled Model Intercomparison Project (CMIP) Phase 6[9] library, and he described the potential value of pretraining digital twins to regenerate missing detail in between sparsely stored checkpoints at short intervals. Other challenges for the use of digital twins include the lack of available specialists to train ML

---

[9] To learn more about CMIP 6, see https://wcrp-cmip.org/cmip-phase-6-cmip6, accessed April 3, 2023.

models. Understanding the extrapolation boundary of current data-driven weather models is also important, as is finding ways to reproducibly and reliably achieve long-term stability despite inevitable imperfections. He concluded that digital twins offer a useful interface between predictions and the stakeholders who could leverage them.

Omar Ghattas, The University of Texas at Austin, described the continuous two-way flow of information between a physical system and a digital twin in the form of sensor, observational, and experimental data. These data are assimilated into the digital twin, and optimal decisions (i.e., control and experimental design) flow from the digital twin to the physical system. In such a tightly coupled system, if stable components are assembled incorrectly, an unstable procedure could emerge. He also explained that although real-time operation is not relevant to climate scales, it is relevant for climate-related issues (e.g., decisions about deploying firefighters to mitigate wildfires). Digital twins are built for high-consequence decision-making about critical systems, which demand uncertainty quantification (e.g., Bayesian data assimilation, stochastic optimal control), he continued, and real-time and uncertainty quantification settings demand reduced-order/surrogate models that are predictive over changing parameter, state, and decision spaces—all of which present massive challenges.

Abhinav Saxena, GE Research, depicted the three pillars of sustainability—environmental protection, economic viability, and social equity. Understanding how the environment and the climate are evolving in combination with how human behaviors are changing and how energy is being used impacts how critical infrastructure could be sustained. He noted that energy generation systems (e.g., gas and wind turbines) and other assets (e.g., engines that consume energy and produce gases) require detailed modeling to be operated more sustainably and efficiently; better weather and climate models are also needed to decarbonize the grid with carbon-free energy generation and microgrid optimization and to achieve energy efficient operations via energy optimization and resilient operation. He asserted that digital twins could

APPENDIX C

help make decades-old systems that experience severe degradation and multiple anomalies more resilient. In this context of life-cycle sustainment, digital twins have the potential to guarantee reliability, optimize maintenance and operations, reduce waste and maximize part life, and reduce costs. He summarized that because physical engineering systems and Earth systems interact with each other, their corresponding digital twins should work in conjunction to best reflect the behaviors of these physical systems and subsequently optimize operations aided by these digital twins toward sustainability.

Elizabeth A. Barnes, Colorado State University, stated that because duplicating the complexity of the Earth system will never be possible, the term "digital twin" is misleading. With the explosion of ML, questions arise about the extent to which a digital twin would be composed of physical theory, numerical integration, and/or ML approximations. Although achieving a "true digital twin" is unlikely, she explained that any future Earth system model will have information (e.g., observations) as an input leading to a target output (e.g., prediction, detection, discovery). She asserted that model developers should be clear about a model's intended purpose, and explainability (i.e., understanding the steps from input to output) is an essential component of that model's success. Explainability influences trust in predictions, which affects whether and how these tools could be used more broadly, allows for fine tuning and optimization, and promotes learning new science. She expressed her excitement about ML approaches that could integrate complex human behavior into models of the Earth system.

Discussion

Incorporating questions from workshop participants, Julianne Chung, Emory University, and Jim Kinter, George Mason University, moderated a discussion among the five panelists. Chung and Kinter asked about the reliability of digital twins in capturing processes and system interactions that vary across scales, as well as about how digital twin results should be communicated to different audiences given uncertainties for decision-making. Saxena replied that reliability depends on a model's level of fidelity; the digital twin should continuously interact with the physical system, learn, and adapt. Trust and explainability are essential to use the prediction from the digital twin to operate the physical system, he added. Ghattas said that the community should not strive to achieve accuracy across all fields and scales. He urged researchers to consider what they want to predict, use a Bayesian framework to infer the uncertainties of the models, and equip the predictions with uncertainty systematically and rationally. Bayesian model selection enables one to attribute probabilities that different models are consistent with the data in meaningful quantities for prediction. Barnes noted that as models become more complex, considering how to assess whether the digital twin is reliable will be critical. Pritchard mentioned the opportunity to use scorecards from Numerical Weather Prediction Centers that provide clear metrics and encouraged further work to develop the right scorecard related to CMIP simulation details. Schneider concurred that developing metrics to assess the quality of a climate model is important; however, he explained that successful weather prediction does not guarantee successful climate prediction because the problems are very different. Because stakeholders' needs vary, no "best" communication strategy exists: a hierarchy of models (e.g., hazard models, catastrophe models) that could better assess risk would be useful for decision-making.

Chung wondered whether uncertainty quantification and calibration are well-posed problems for Earth system simulation. Schneider described them as ill-posed problems because the number of degrees of freedom that are unresolved far exceeds the number of degrees of freedom available in the data. Therefore, additional prior information (e.g., governing equations of physics and conservation laws, domain-specific knowledge) is needed to reduce the demands on the data. Ghattas added that most inverse problems are ill-posed. The data are informative about the parameters in the low-dimensional manifold, and the other dimensions are handled via a regularization operator; for example, the Bayesian framework allows one to bring in prior knowledge to fill the gaps. He stressed that Bayesian inversion is extremely challenging on large-scale ill-posed problems and reduced-order models and surrogates would be useful.

Kinter asked the panelists to share examples of useful digital twins with current model fidelity and computational resources. Schneider mentioned that climate models have been useful in predicting the global mean temperature increase. A demand for zip code–level climate information exists, but current models are not fit for that purpose. Barnes described progress in identifying sources of predictability in the Earth system. She asserted that relevant information from imperfect climate models is vital for understanding how the real world will behave in the future. Saxena explained that when GE wants to install a new wind farm, models provide a sufficient level of detail to determine how to protect assets from weather and environmental impacts. Pritchard reiterated that digital twins enable massive ensembles, creating an opportunity to study the tail statistics of rare events and to assimilate new data streams.

PANEL 3: TRANSLATION OF PROMISING PRACTICES TO OTHER FIELDS

During the third panel, workshop participants heard brief presentations from experts in polar climate, AI algorithms, ocean science, carbon cycle science, and applied mathematics; they discussed how digital twins could be useful in their research areas and where digital twins could have the greatest future impacts.

Cecilia Bitz, University of Washington, championed the use of digital twins to understand Earth system components such as sea ice. By increasing the realism of the ocean and atmosphere and achieving improvements in those components, improvements in downstream components could be realized. Highlighting opportunities for advances in sea ice components, she referenced ECMWF's high-resolution simulation of sea ice—with kilometer-scale flows of sea ice in the horizontal and large openings dynamically occurring—as an example of the progress enabled by moving to high resolution to view dynamic features. She expressed interest in broadening the kind of physics in the digital twin framework as well as developing parameterizations to be more efficient, which is not only a significant challenge but also a great opportunity to expand the capabilities of the surface component.

Anima Anandkumar, California Institute of Technology/NVIDIA, discussed efforts to incorporate fine-scale features in climate simulations using neural operators. She explained that ML with standard frameworks captures only finite dimensional information; however, neural operators (which are designed to learn mappings between function spaces) enable finer scales, evaluate throughout the domain, and capture physics beyond the training data. Training and testing can be conducted at different resolutions, and constraints can be incorporated into the training. She encouraged using data-driven approaches to learn from a wealth of historical weather data and to incorporate extreme weather events into models—these approaches encompass a wide range of fields and new possibilities for generative AI in science and engineering.

Emanuele Di Lorenzo, Brown University, described work to engage coastal stakeholders and researchers to co-design strategies for coastal adaptation and resilience in Georgia. Supported by a community-driven effort, the Coastal Equity and Resilience (CEAR) Hub's[10] initial modeling and forecasting capabilities provide water-level information at the scale where people live. A network of ~65 sensors that are distributed and interconnected wirelessly along the coast around critical infrastructure provides actionable data that are important for decision-makers, which are streamed into dashboards that are continuously redesigned with their input. He explained that as the focus on equity and climate justice expands, new factors related to resilience are emerging (e.g., health and socioeconomic well-being) that demand new community-driven metrics. The CEAR Hub plans to expand its sensor network to measure air and water quality and urban heating and combine these new sources of data with social data to develop the metrics.

Anna Michalak, Carnegie Institution for Science, explained that the carbon cycle science community focuses on questions around quantification (i.e., how greenhouse gas emissions and uptake vary geographically at different resolutions as well as their variability over time), attribution (i.e., constraining the processes that drive the variability seen in space and time, which

---

[10] The website for the CEAR Hub is https://www.cearhub.org, accessed March 6, 2023.

APPENDIX C

requires a mechanistic-level understanding), and prediction (e.g., global system impact if emissions hold a particular trajectory or if climate changes in a particular way). These questions are difficult because carbon cycle scientists work in a data-poor environment—both in situ and remote sensing observations are sparse in space and time, and fluxes cannot be measured directly beyond the kilometer scale. The community has moved toward the use of ensembles of models as well as ensembles of ensembles to confront this problem. She said that the current best strategy to address the uncertainty associated with quantification, attribution, and prediction is to run multiple simulations of multiple models, with the whole community working on different incarnations of these models. She also suggested simplifying models before increasing their complexity to understand fundamental mechanistic relationships. The "holy grail" for digital twins in carbon cycle science, she continued, would be to use them for hypothesis testing, to diagnose extreme events, and for prediction. .

John Harlim, The Pennsylvania State University, defined digital twins as a combination of data-driven computation and modeling that could involve data simulation, modeling from first principles, and ML algorithms. He emphasized that the fundamental success of digital twins depends on their ability to compensate for the modeling error that causes incompatibility of numerical weather prediction models and climate prediction models. If one could compensate appropriately for this modeling error, the same model proven to be optimal for state estimation could also accurately predict climatological statistics. However, he asserted that this is not achievable for real-world applications. Specific domain knowledge is critical to narrow the hypothesis space of models, which would help the ML algorithm find the underlying mechanisms. Using a model-free approach, he continued, is analogous to allowing the algorithm to find a solution from a very large hypothesis space of models. Such a practice could reduce the bias, but this bias has to be balanced with the variance error. He stressed that the success of digital twins in other fields depends on whether enough informative training data exist to conduct reliable estimations.

Discussion

Incorporating questions from workshop participants, Leung and Chung moderated a discussion among the five panelists. Leung wondered how limited the field is by data when powerful new techniques could be leveraged to extend beyond training data. Michalak replied that a process-based, mechanistic understanding is needed to anticipate future climate system evolution. She said that new modeling techniques could be used to better leverage limited observations, which could assist with uncertainty quantification; however, these new approaches would not fundamentally change the information content for existing observations. She emphasized that new tools offer promise but are not a panacea across use cases and disciplines. Anandkumar elaborated on the ability of these new tools to extrapolate beyond training data. She said that neural operators are being used as surrogates for solving partial differential equations (PDEs) and other equations that can be embedded; at the same time, data could be combined with physics for nonlinear representations. Michalak added that this is feasible only if the challenge is on the PDE side, not if the challenge relates to the parametric and structural uncertainties in the models.

Chung asked how uncertainties could be better understood, quantified, and communicated. Anandkumar responded that ML has great potential, although it is still an emerging approach; with its increased speed, thousands of ensemble members could be created—having this many ensemble members and the ability to incorporate uncertainties is critical. The next step could be to use emerging frameworks such as diffusion models, as long as they incorporate uncertainty accurately. Harlim noted that developing a digital twin that predicts the response of second-order statistics would be very difficult, especially when the system is spatially extended and non-homogeneous. He noted that the ensemble mean is widely accepted to provide accurate predictions; however, a question remains about whether co-variants provide uncertainty about the estimations.

Leung inquired about strategies to work with decision-makers to define the requirements of digital twins. Di Lorenzo advocated for co-designed, community-driven research projects that use a transdisciplinary approach. Meeting with project stakeholders and community leaders raises awareness, increases engagement, and creates ownership; the scientist's role is to provide support to articulate the problem posed by the stakeholders. To initiate such projects, he said that scientists should identify boundary organizations with existing ties to the community. Bitz described her work with Indigenous communities in Alaska to better understand the threats of coastal erosion, which prioritizes listening to their concerns and building trusted relationships. She urged scientists to use their knowledge to engage in problems directly and to collaborate with scientists from other domains. Michalak suggested fostering relationships with the private sector to ensure that its investments in climate solutions have the maximum possible impact.

Leung posed a question about the difference between digital twins and the modeling that has been ongoing since the 1960s. Di Lorenzo noted that digital twins are tools with applications for decision-making in the broader community rather than just the scientific community. If the digital twin is meant to serve the broader community, however, he said that the term "digital twin" is too confusing. Anandkumar observed that digital twins are data-driven, whereas modeling over the past decades has primarily used data for calibration. Leung also wondered how social science data could inform digital twin studies. Bitz explained that ground truthing data with local community knowledge is a key part of model development, and social scientists could facilitate that process. Di Lorenzo urged researchers to include social dimensions in digital twin platforms, as thinking only about the physical world is an obsolete approach.

PANEL 4: DISCUSSION ON TRANSPARENCY, SOCIETAL BENEFIT, AND EQUITY

Incorporating questions from workshop participants, Ye and Parris moderated the workshop's final discussion among three experts on transparency, societal benefit, and equity considerations for the use of digital twins: Amy McGovern, University of Oklahoma; Mike Goodchild, University of California, Santa Barbara; and Mark Asch, Université de Picardie Jules Verne. Ye invited the panelists to define digital twins. McGovern described digital twins as high-fidelity copies such that the digital twin and the original observations are as indistinguishable as possible and allow exploration of what-if scenarios. She added that a human dimension is crucial for the digital twin to enable decision-making. Focusing on the capacity for "high resolution," Goodchild commented that a question arises about what threshold has been passed, as resolution has become finer and will continue to become finer. However, he noted that the human, decision-making, and sustainability components are defining characteristics of digital twins. Asch underscored the need to educate decision-makers in how to use the tools that scientists develop; much work remains to communicate that digital twins are decision-making tools, not "magic wands." Parris inquired about how a digital twin extends beyond modeling. Goodchild highlighted the use of digital twins to visualize fine-resolution geospatial data, which will appeal to a broad audience, although visualizing uncertainty in these data is very difficult. McGovern explained that today's modeling world includes data in different formats and scales, with and without documentation—a digital twin could provide a consistent form to access these data, which could enable AI and visualization. Asch urged more attention toward improving the modeling of uncertainty, especially in light of recent advances in computational power. He emphasized that decisions derived from digital twins are probabilistic, based on the relationship between value and risk, not deterministic. In response to a question from Ye about unique approaches to visualize uncertainty, McGovern described an initiative where those creating visualizations are conducting interviews with end users to understand how uncertainty affects their trust in the model. A next step in the project is allowing the end users to manipulate the underlying data to better understand uncertainty.

Parris asked the panelists to share examples of digital twins that provide societal benefit. Goodchild described the late 1990s concept of the Digital Earth as a "prime mover" in this space and noted that the literature over the past 30 years includes many examples of interfaces between "digital twins" and the decision-making process, especially in industry. McGovern said

APPENDIX C

that DestinE could allow people to explore how climate and weather will directly impact them; however, she stressed that more work remains for DestinE to reach its full potential. Asch suggested drawing from the social sciences, humanities, and political sciences to help quantify qualitative information. Integrating more diverse beliefs and values into science is critical, he continued, although enabling cross-disciplinary collaboration is difficult within existing funding streams. In response to a question from Parris about how digital twins integrate natural and human systems, Asch described work in the Philippines to model the spread of viral epidemics. He noted that creating dashboards is an effective way for end users to interact with a complicated problem; however, more work remains to model social and psychological phenomena. Goodchild highlighted the value of understanding interactions between humans and their environment in terms of attitudes and perceptions, and he referenced the National Science Foundation's coupled natural and human systems program and others' successes with agent-based modeling, where the behavior of humans is modeled through a set of rules.

Ye posed a question about the trade-offs of using data in a digital twin and maintaining privacy. Goodchild remarked that location privacy is becoming problematic in the United States. Location data are of commercial interest, which makes it even more difficult to impose regulations. He posited that the issue of buying and selling location data without individuals' awareness should be given more attention. With finer resolution, these ethical issues become critical for digital twins, and he suggested implementing a regulation similar to the Health Insurance Portability and Accountability Act (HIPAA). McGovern acknowledged the disadvantages of location data but also highlighted the advantages. She noted the need to evaluate trade-offs carefully to improve models while protecting privacy; a HIPAA-like regulation could make it difficult to obtain useful information for digital twins. Asch commented that the potential for abuse is significant, but high-level data-sharing agreements and data security technology could address this issue.

Parris inquired about the role of bias in digital twins. McGovern observed that although bias might be helpful, most often it is harmful and should be discussed more often in relation to weather and climate data. The first step is recognizing that bias exists at all stages of digital twin creation, and that systemic and historical biases affect which data are missing. Goodchild added that limiting the spatial extent of digital twins could help to address this issue, and Asch proposed that critical reasoning be used to detect bias. Given these issues, Ye asked how to improve confidence in digital twins. Asch stressed that transparency and reproducibility are key to increasing digital twin acceptance, and users should be able to follow a digital twin's reasoning as well as understand how to use and exploit it. McGovern stated that co-development with the end user helps advance both confidence and trustworthiness. Goodchild explained that some uncertainty in what digital twins predict and how they operate will always exist. He said that ethical issues related to digital twin reusability could also arise, and enforcing fitness for use is essential; "repurposing" is a significant problem for the software industry to confront.

Parris posed a question about strategies to build a diverse community of researchers for digital twins. Goodchild suggested first identifying the subsets of problems for which a digital twin could be useful. Asch said that understanding the principles of modeling and the problem-solving process for digital twins is key. He reiterated the value of bringing philosophy and science together to better develop and use these tools, emphasizing reasoning as a means to help democratize them. He also encouraged increasing engagement with students in developing countries. McGovern stressed that, instead of waiting for students to enter the pipeline, the current workforce should be given meaningful problems to solve as well as training on AI methods; furthermore, offering AI certificate programs at community colleges plays an important role in creating a more diverse workforce.

DISCLAIMER This Proceedings of a Workshop—in Brief was prepared by **Linda Casola** as a factual summary of what occurred at the workshop. The statements made are those of the rapporteur or individual workshop participants and do not necessarily represent the views of all workshop participants; the planning committee; or the National Academies of Sciences, Engineering, and Medicine.

COMMITTEE ON FOUNDATIONAL RESEARCH GAPS AND FUTURE DIRECTIONS FOR DIGITAL TWINS Karen Willcox (Chair), Oden Institute for Computational Engineering and Sciences, The University of Texas at Austin; **Derek Bingham**, Simon Fraser University; **Caroline Chung**, MD Anderson Cancer Center; *Julianne Chung*, Emory University; **Carolina Cruz-Neira**, University of Central Florida; **Conrad J. Grant**, Johns Hopkins University Applied Physics Laboratory; *James L. Kinter III*, George Mason University; **Ruby Leung**, Pacific Northwest National Laboratory; **Parviz Moin**, Stanford University; **Lucila Ohno-Machado**, Yale University; **Colin J. Parris**, General Electric; **Irene Qualters**, Los Alamos National Laboratory; **Ines Thiele**, National University of Ireland, Galway; **Conrad Tucker**, Carnegie Mellon University; **Rebecca Willett**, University of Chicago; and **Xinyue Ye**, Texas A&M University–College Station. * Italic indicates workshop planning committee member.

REVIEWERS To ensure that it meets institutional standards for quality and objectivity, this Proceedings of a Workshop—in Brief was reviewed by **Jeffrey Anderson**, National Center for Atmospheric Research; **Bryan Bunnell**, National Academies of Sciences, Engineering, and Medicine; and **Xinyue Ye**, Texas A&M University-College Station. **Katiria Ortiz**, National Academies of Sciences, Engineering, and Medicine, served as the review coordinator.

STAFF Patricia Razafindrambinina, Associate Program Officer, Board on Atmospheric Sciences and Climate, *Workshop Director*; **Brittany Segundo**, Program Officer, Board on Mathematical Sciences and Analytics (BMSA), *Study Director*; **Kavita Berger**, Director, Board on Life Sciences; **Beth Cady**, Senior Program Officer, National Academy of Engineering; **Jon Eisenberg**, Director, Computer Science and Telecommunications Board (CSTB); **Samantha Koretsky**, Research Assistant, BMSA; **Tho Nguyen**, Senior Program Officer, CSTB; **Michelle Schwalbe**, Director, National Materials and Manufacturing Board (NMMB) and BMSA; **Erik B. Svedberg**, Senior Program Officer, NMMB; and **Nneka Udeagbala**, Associate Program Officer, CSTB.

SPONSORS This project was supported by Contract FA9550-22-1-0535 with the Department of Defense (Air Force Office of Scientific Research and Defense Advanced Research Projects Agency), Award Number DE-SC0022878 with the Department of Energy, Award HHSN263201800029I with the National Institutes of Health, and Award AWD-001543 with the National Science Foundation.

This material is based on work supported by the U.S. Department of Energy, Office of Science, Office of Advanced Scientific Computing Research, and Office of Biological and Environmental Research.

This project has been funded in part with federal funds from the National Cancer Institute, National Institute of Biomedical Imaging and Bioengineering, National Library of Medicine, and Office of Data Science Strategy from the National Institutes of Health, Department of Health and Human Services.

Any opinions, findings, conclusions, or recommendations expressed do not necessarily reflect the views of the National Science Foundation.

This proceedings was prepared as an account of work sponsored by an agency of the United States Government. Neither the United States Government nor any agency thereof, nor any of their employees, makes any warranty, express or implied, or assumes any legal liability or responsibility for the accuracy, completeness, or usefulness of any information, apparatus, product, or process disclosed, or represents that its use would not infringe privately owned rights. Reference herein to any specific commercial product, process, or service by trade name, trademark, manufacturer, or otherwise does not necessarily constitute or imply its endorsement, recommendation, or favoring by the United States Government or any agency thereof. The views and opinions of authors expressed herein do not necessarily state or reflect those of the United States Government or any agency thereof.

SUGGESTED CITATION National Academies of Sciences, Engineering, and Medicine. 2023. *Opportunities and Challenges for Digital Twins in Atmospheric and Climate Sciences: Proceedings of a Workshop—in Brief*. Washington, DC: The National Academies Press. https://doi.org/10.17226/26921.

Division on Engineering and Physical Sciences

Copyright 2023 by the National Academy of Sciences. All rights reserved.

NATIONAL ACADEMIES Sciences Engineering Medicine

The National Academies provide independent, trustworthy advice that advances solutions to society's most complex challenges.

www.nationalacademies.org

D

Opportunities and Challenges for Digital Twins in Biomedical Research: Proceedings of a Workshop—in Brief

Opportunities and Challenges for Digital Twins in Biomedical Research: Proceedings of a Workshop—in Brief (National Academies of Sciences, Engineering, and Medicine, The National Academies Press, Washington, DC, 2023) is reprinted here in its entirety. The original Proceedings of a Workshop—in Brief is available at https://doi.org/10.17226/26922.

Proceedings of a Workshop—in Brief

Opportunities and Challenges for Digital Twins in Biomedical Research

Proceedings of a Workshop—in Brief

The digital twin (DT) is an emerging technology that builds on the convergence of computer science, mathematics, engineering, and the life sciences. Given the multiscale nature of biological structures and their environment, biomedical DTs can represent molecules, cells, tissues, organs, systems, patients, and populations and can include aspects from across the modeling and simulation ecosystem. DTs have the potential advance biomedical research with applications for personalized medicine, pharmaceutical development, and clinical trials.

On January 30, 2023, the National Academies of Sciences, Engineering, and Medicine hosted a public, virtual workshop to discuss the definitions and taxonomy of DTs within the biomedical field, current methods and promising practices for DT development and use as various levels of complexity, key technical challenges and opportunities in the near and long term for DT development and use, and opportunities for translation of promising practices from other field and domains. Workshop panelists highlighted key challenges and opportunities for medical DTs at varying scales, including the varied visions and challenges for DTs, the trade-offs between embracing or simplifying complexity in DTs, the unique spatial and temporal considerations that arise, the diversity of models and data being used in DTs, the challenges with connecting data and models across scales, and implementation issues surrounding data privacy in DTs. The first in a three-part series, this information-gathering workshop will inform a National Academies consensus study on research gaps and future directions to advance the mathematical, statistical, and computational foundations of DTs in applications across science, medicine, engineering, and society.[1]

VISIONS AND CHALLENGES FOR DIGITAL TWINS

Reinhard Laubenbacher, University of Florida, the plenary session's first speaker, described the DT as a computational model that represents a physical system; a data stream between the system and the model is used to recalibrate the model periodically so that it continues to represent the system as it changes over the lifespan of its operation. He stated that DTs are revolutionizing industry (e.g., aeronautics) with applications including forecasting and preventative maintenance. He proposed that this notion of preventive maintenance could be reflected in medical DTs, with the potential for patients to avoid unscheduled visits to the doctor. Medical DTs

---

[1] To learn more about the study and to watch videos of the workshop presentations, see https://www.nationalacademies.org/our-work/foundational-research-gaps-and-future-directions-for-digital-twins, accessed February 10, 2023.

APPENDIX D

combine models of human biology with operational data to make predictions and optimize therapy. However, he cautioned that medical DTs are very different from those used in industry. The medical field leverages different types of data, which are often sparse. With this somewhat limited knowledge about humans, human biology has to be reverse-engineered and encoded in a model. Thus, experimental data (often from mice or tissue cultures) is frequently used in building the computational models for medical DTs. If implemented successfully, he indicated that medical DTs could eventually lead to a paradigm shift from curative to preventive medicine as well as help to personalize medical interventions, improve decision-support capabilities for clinicians, develop drugs more quickly at a lower cost, and discover personalized therapies.

Laubenbacher emphasized that because medical DTs are much more difficult to build than computational models, distinguishing appropriate use cases for a DT versus a model is critical. For example, to improve type 1 diabetes care, an artificial pancreas[2] can be linked to a sensor in a patient's bloodstream that continually measures glucose, which is connected to a closed-loop controller that determines how much insulin the patient's pump should inject. Because this model recalibrates itself based on the patient's needs, he noted that this device is a medical DT. To improve cardiac care, a three-dimensional image of a patient's heart and computational fluid dynamics could be used to simulate blood flow to make predictions about the patient's risk for a blood clot or to make decisions about bypass surgery. The classification of this technology depends on its use—if used once for decision support, it is a personalized model; if calculations for risk and medication are updated periodically, it is a medical DT.

Laubenbacher presented several challenges associated with building a medical DT: (1) identifying and solving difficult scientific problems that arise at different scales (e.g., systems biology, biophysics, the immune system); (2) addressing gaps in modeling (e.g., multiscale hybrid stochastic models, model design that facilitates updates and expansion, reusable models, and model standards);

(3) developing appropriate collection modalities for patient data (e.g., noninvasive technologies and imaging capabilities); (4) developing novel forecasting methods (i.e., learning from successful hurricane forecasting); (5) developing data analytics methods for model recalibration from patient measurements; (6) training a highly educated workforce; and (7) creating appropriate funding models for individual medical DT projects from conception to prototype, and for larger infrastructure development projects.

Karissa Sanbonmatsu, Los Alamos National Laboratory, the plenary session's second speaker, highlighted several differences between DTs and computational models, including that DTs should provide real-time information and insights and can be used across the lifetime of a specific asset for decision-making. DTs can be classified as component twins (i.e., modeling a single part of a system), asset twins (i.e., modeling multiple parts of a system), system twins (i.e., modeling the whole system), or process twins (i.e., focusing on how parts of a system work together).

Sanbonmatsu pointed out that in the field of integrative biology, computational methods are used to integrate many types of experimental data into a coherent structural model. Computational modeling in biology is conducted at many scales—for example, cellular and molecular scale (e.g., molecular dynamics simulations); organ, tumor, and microenvironment scale (e.g., Bayesian modeling); whole-human, multisystem, and population scale (e.g., epidemiological modeling); and multiple scales. Molecular simulations, in particular, are used to understand molecular machines and ion channels, to determine structural models, to design better drugs and vaccines, and to advance applications in biotechnology. She described four key challenges in the simulation of biomolecules: (1) many biological systems are inherently discrete, where phenomena at the atomistic level have important macroscopic implications for cell fate and disease; (2) length scales are approximately 10 orders of magnitude, and time scales are approximately 20 orders of magnitude; (3) information matters, especially the sequence of bases in the human genome; and (4) systems are highly

---

[2] The website for Tandem Diabetes Care is https://www.tandemdiabetes.com, accessed February 27, 2023.

charged, which requires long-range electrostatic force calculations. Ultimately, such simulations require many time steps to be relevant; until supercomputers have the necessary capabilities, she continued, researchers will continue to leverage experimental data for multiscale integrative biology.

Discussion

Incorporating questions from workshop participants, Irene Qualters, Los Alamos National Laboratory, moderated a discussion with the plenary session speakers. She asked if standards would help to make DTs more rigorous and trustworthy in the biomedical community. Laubenbacher replied that the industrial space uses Modelica, a standard for building models of modular components that can be integrated for DTs. The Systems Biology Markup Language, which applies to differential equation models, is an established standard in the biomedical community; however, he noted that much more work remains in the development of standards overall for medical DTs to be used at scale.

Qualters wondered whether successful examples of DTs at the cellular and molecular scale exist. Sanbonmatsu responded that although several successful simulations exist and efforts are under way to develop real-time molecular models, the community does not yet have the capability to work in real time via DTs. Laubenbacher pointed out that future DTs would have to cross scales, because most drugs work at the cellular level but have effects at the tissue level. Qualters also posed a question about whether a DT could represent a population instead of only an individual—for example, a disease model that is continually updated with new trial data. Laubenbacher observed that EpiSimS, an agent-based simulation model, captures an entire city based on details that represent the population. He cautioned that all key characteristics of a population should be included in a medical DT, as sampling the parameter space is insufficient.

In response to a question, Sanbonmatsu remarked that a medical DT should track a person's full medical history, including drug interactions and health conditions. Because all humans are different, the sequence of each person's chromosomes would also be useful for the model. Laubenbacher added that collecting the right data (without invasive procedures for patients) and determining how those data influence parameter updates is a significant challenge in building medical DTs. Qualters inquired about the role of organs-on-a-chip technology in building higher-fidelity digital models. Sanbonmatsu noted that such technology will be critical in bridging the gap between the actual patient and the petri dish to build computational models of an organ. Laubenbacher commented that because so little can be measured directly in humans, these technologies are beneficial, as are good cell culture systems for primary tissue. He asserted that collecting information about humans without using mice and monkeys is a step forward.

EMBRACING OR SIMPLIFYING COMPLEXITY IN DIGITAL TWINS

During the first panel, workshop participants heard brief presentations on DTs at the cellular and molecular scale from the perspectives of pharmaceutical development, clinical practice, and research. The challenge of embracing or simplifying complexity in DTs was raised.

Jeffrey Sachs, Merck & Co., Inc., defined a DT as a simulation with sufficient fidelity for an intended purpose. This simulation could represent processes, objects, or physical systems; include visualization, artificial intelligence (AI), or machine learning (ML); and involve real-time sensors. He further described DTs as a type of modeling and simulation that is personalized to a person or a population.

Sachs provided an overview of how pharmaceutical companies use modeling and simulation in clinical trials and drug development. For example, to rank order the potential efficacy of a drug or vaccine, ab initio modeling is used, with input from physics and chemistry. To determine potential drug interactions, quantitative systems pharmacology and physiologically based pharmacokinetics techniques are leveraged, with model input from measurements of biological and chemical assays in experiments. To understand which drug infusion speed is best for an individual, population pharmacokinetic-pharmacodynamics and quantitative

APPENDIX D

systems pharmacology techniques are used, with model input from both large historical data sets of biological and chemical assays and data from the individual—a type of modeling often required by regulatory agencies.

Sachs highlighted several challenges related to assessing uncertainty in modeling. Data context matters, and one should be able to measure the right thing in the right way. Model complexity and identifiability as well as computability and data heterogeneity also present obstacles. Furthermore, a question remains about when summary results from one model should be used as input to the next-level model.

Mikael Benson, Karolinska Institute, observed that medication is often ineffective for patients with complex, malignant diseases, leading to both personal suffering and financial strain. Each disease might involve thousands of genes across multiple cell types, which can vary among patients with the same diagnosis. With high-resolution models to identify the optimal drug for each patient, DTs could bridge this gap between disease complexity and health care options. He described the Swedish Digital Twin Consortium's[3] approach: a DT of a single patient is created based on the integration of detailed clinical and high-resolution data. The DT is then computationally treated with thousands of drugs to determine which offers a cure, and the best drug is given to the patient (Figure 1). He mentioned that this approach has worked for animal models, patient cells, and clinical data analysis.

Benson pointed out that the complexity of the data and analysis required for medical DTs could disrupt medical practice, research, and publishing. DTs could unsettle explainable medical practice, as diagnoses that used to be based on a few blood tests and images would be based on mechanisms that involve thousands of genes. However, examples exist in which complex clinical data can be reduced to lower dimensions. Similarly, although traditional medical research is based on a detailed analysis of a limited number of variables, medical DT research is based on an analysis of thousands of genes across multiple cell types in mouse models through

[3] The Swedish Digital Twin Consortium website is https://sdtc.se, accessed February 27, 2023.

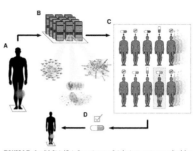

FIGURE 1 The Swedish Digital Twin Consortium uses digital twins to create personalized drug prescriptions based on simulating different interventions on the digital twin of the patient.
SOURCE: Mikael Benson, Karolinska Institute, presentation to the workshop, from Björnsson, B., Borrebaeck, C., Elander, N. et al., 2020, "Digital Twins to Personalize Medicine," *Genome Medicine* 12:4, https://doi.org/10.1186/s13073-019-0701-3, Copyright 2019, https://creativecommons.org/licenses/by/4.0.

clinical studies. Each part of the study is interdependent, and one single part cannot be extracted for a manuscript without losing context. Each part contains highly complicated data, and results are difficult to validate with conventional methods; validation often depends on reproducibility studies. He asserted that journal editors should address the fact that current manuscript format and peer-review limitations do not accommodate the multilayered and multidisciplinary nature of DT studies.

Juan Perilla, University of Delaware, described the study of retroviruses at the molecular level using computational models that can be interrogated to predict new properties. When this work began more than a decade ago, knowledge about cytoplasmic events was limited. Over time, those previously obscure events were revealed to be crucial in understanding how viruses take advantage of cells.

Perilla provided an overview of the process used to understand whether the HIV-1 capsid could translocate into the nucleus of a cell. To address this question, he and his team modeled biological parts and developed novel techniques to perturb the system mechanically and observed how it restored, while ensuring that the model was reproducing reality. By introducing other known chemical aspects and probing elastic properties, they could visualize the process not being captured by experimentalists, which allowed them to make predictions, test the model, and obtain specific chemical information about the process. Understanding the system's physical and mechanical properties confirmed that the HIV-1 capsid can translocate to the nucleus; this knowledge also enables future development of new materials.

Rommie Amaro, University of California, San Diego, described her work in developing molecular simulations at the mesoscale, with a focus on building and understanding viruses and the systems with which they interact. She and her team are creating atomic-level models of biological systems, using high-performance simulations to bring together and interrogate different types of experimental data sets and to view the trajectory of atomic-level dynamics over time.

Amaro explained that these types of simulations make it possible to predict the dynamics of structures that are otherwise invisible experimentally, thus augmenting what experimentalists cannot see (e.g., the glycan shield). Therefore, the simulations work in tandem with interrogation by experiment, she continued, and these methods enable the development of predictive models from the nanoscale to the microscale. Work is ongoing to model the dynamics of single viruses (e.g., SARS-CoV-2) using large-scale supercomputing architectures, which is crucial for vaccine development and drug design. These simulations are also being used to study airborne pathogens inside respiratory aerosols. Because DTs can be used to connect the molecular scale to population-level outcomes to make predictions, she championed Benson's assertion that interdisciplinary peer reviewers are essential to move the field forward.

Discussion

Incorporating questions from workshop participants, Ines Thiele, National University of Ireland, Galway, moderated a discussion among the four panelists. She posed a question about strategies to validate predictions from DT simulations when experimental data cannot be obtained. Amaro explained that many indirect experiments can validate emerging predictions. When hypotheses and results can be interrogated and replicated by other research groups worldwide using the same or similar models, she asserted that trust will increase among experimentalists. Perilla suggested leveraging negative controls for validation. Sachs added that building trust relates to the application, and validation becomes especially critical in health care, where replication is only the first step in the process. Offering a clinical perspective, Benson remarked that testing a prediction in a clinical setting relates to diagnostics, sensitivity, and specificity for computational treatment with thousands of drugs. He explained that precision and recall are key for known drugs, and clinical trials for new drugs have standard measures of clinical improvement.

In response to a question about statistical considerations for DTs, Sachs commented that understanding the distribution of parameter values, characteristics, and outcomes is essential if DTs are to be used safely, as variability impacts efficacy.

Thiele asked when highly detailed precision simulations should be used instead of high-level surrogate models. Perilla explained that the study of HIV primarily focuses on cell components; the interactions of interest depend on a specific three-letter sequence and detail, which requires the incorporation of atomistic knowledge. When physical properties of a system arise from chemical properties, reducing the complexity of the system would have a negative impact. For example, although all retroviruses look alike, he pointed out that they are extremely different, and reducing the complexity of the model is dangerous. Amaro indicated that model selection depends on the domain of interest as well as what one is trying to predict. To design novel chemical matter, one has to keep track of all of the atoms in the design; however, for larger-level effects, one can coarse-grain away the most computationally intensive aspects of a calculation. She noted that over time and for particular systems, decision-making to determine what can be coarse-grained away will become more straightforward.

SPATIAL AND TEMPORAL CONSIDERATIONS IN DIGITAL TWINS

During the second panel, workshop participants heard brief presentations on DTs at the organ, tumor, and microenvironment scale in the areas of computational oncology and ophthalmologic data sciences. A common theme emerged of addressing spatial and temporal considerations in DTs.

Tom Yankeelov, The University of Texas at Austin, emphasized the need to calibrate mechanism-based models with patient-specific data to make patient-specific predictions. He mentioned that tumor forecasts are improving in terms of accuracy, but like weather forecasts, tumor forecasts are not DTs. Promising applications for the use of DTs include linking a mechanism-based model of drug delivery to a model of tumor response, thereby enabling in silico clinical trials where the drug dose and schedule could be systematically varied. Because the optimal regimen that emerges for an individual patient is often different from the standard population-based regimen, he asserted that mechanism-based mathematical modeling is essential.

Yankeelov highlighted three key mathematical, computational, and data challenges in the development of medical DTs. First, models that capture relevant biology and that can be calibrated with available data are needed, as are models that are actionable. Second, because calibration, model selection and updates, and therapy optimization are expensive, strategies to accelerate these loops would be useful. Third, increased access to "smarter" patient data would be beneficial. As diseases and therapies become increasingly specific and complex, he urged researchers to deprioritize AI and big data in favor of mechanism-based models that can be calibrated with patient data. Furthermore, because a patient is unlikely to be cured with only one therapy, he remarked that these models should be updated continually during treatment. To begin to address these challenges, techniques from other fields could be applied such as optimal control theory, reduced order modeling, uncertainty quantification, and data assimilation.

Jayashree Kalpathy-Cramer, University of Colorado Denver, offered a relevant definition of DTs derived from ChatGPT: "a virtual representation of a physical object or system that can be used for simulation, analysis, and monitoring."[4] DTs that use medical imaging could

---
[4] J. Kalpathy-Cramer, University of Colorado Denver, presentation to the workshop, January 30, 2023.

be leveraged to inform and improve patient care via procedure planning, medical education, individual risk prediction, and prediction of cancer drug delivery.

Kalpathy-Cramer discussed a study on optimizing the regimen and dosage of anti-vascular endothelial growth factor (anti-VEGF) therapies for glioblastoma patients. The study mathematically modeled blood flow through the vessels, and three-dimensional virtual voxels generated from optical imaging in an animal model were used as input to the mathematical models. With those models, one could observe the effects of the anti-VEGF therapy on the vasculature, and what was seen in the mathematical models and simulations turned out to be similar to what was observed in humans. Presenting a completely different approach using data-driven modeling, she depicted work to identify treatment-requiring retinopathy of prematurity. ML and deep learning were used to develop a severity scale for the disease based on available data, and the trajectories were quantified of patients who would require treatment versus those who would not. A so-called DT emerged of the path an individual patient might take throughout their care, including the point at which they either respond to or require more therapy. No mathematics are used to make such predictions. However, she stressed that although data-driven approaches are powerful, they are sometimes entirely wrong. As DTs also have the potential to be both powerful and incorrect, she described a combination of data-driven and mechanistic approaches as the best path forward to make better decisions for patients.

Discussion

Incorporating questions from workshop participants, Caroline Chung, MD Anderson Cancer Center, moderated a discussion among both speakers and two experts in intelligent systems engineering and applied mathematics—James A. Glazier, Indiana University, and Petros Koumoutsakos, Harvard University, respectively. In light of the nuanced differences between computational models and DTs, Chung posed a question about how to determine whether a problem warrants the use of a DT. Yankeelov observed that if a patient has cancer and is prescribed several drugs, the order and delivery of those drugs is critical—a DT would help determine the optimal therapeutic regimen that could be adopted in a clinical trial. Glazier added that because real-time evaluation of a drug's effectiveness is not currently feasible, a DT of a patient could be used as a virtual control to determine whether a drug is working and to tune the therapy accordingly. Koumoutsakos highlighted issues that arise in computational modeling in terms of the mismatch between what can be measured in a medical laboratory and what can be simulated owing to discrepancies in the spatial and temporal scales of the data. Chung asked what types of data are needed to build a DT. Although patient data are sparse, Kalpathy-Cramer stressed the importance of balancing the patient's best interests with the desire to learn more by collecting only the minimum amount of data necessary for a model to be useful both in terms of input and output. She added that a forward model that generates synthetic data could also be useful.

Chung inquired about the role of a human-in-the-loop to intervene and react to a DT. Glazier pointed out that although the aeronautics industry uses a reactive approach (i.e., if a DT predicts an engine failure, the engine is removed from the aircraft and repaired), the biomedical field would instead benefit from continual interventions to maintain a patient's healthy state. In the absence of single protocols, this approach would require significant changes in perspectives on medication. He envisioned a future with human-out-of-the-loop regulation of health states but acknowledged challenges in the loss of autonomy and the possibility for catastrophic failure. Self-driving cars provide an important lesson for the future of medical DTs in terms of the value of bootstrapping, he continued. He suggested rolling out DTs even if they do not predict well or far in time, and creating millions of replicas to understand where they fail and how to make improvements. Mechanisms could then be designed to automatically learn from the failures at the population level to improve model parameters and structures.

Chung asked about the pipelines needed for the consistent and accurate parameterization of DT models. Koumoutsakos replied that the first step is to determine the metric for how closely the model represents reality, incorporating uncertainty. The next step is to combine domain-specific knowledge with system-specific data and to leverage AI, ML, and uncertainty quantification techniques. Once the model is developed, he explained, one should be prepared to do iterations and optimizations to the parameters based on the chosen metric.

Given the dynamic nature of DTs, Chung wondered about the time scale of measurement necessary to advance DTs at the organ, tumor, and microenvironment level. Kalpathy-Cramer said that the time scale is determined by the length of time it takes to see a drug response, and she suggested developing strategies to handle multiple time scales. Yankeelov remarked that daily model optimization and prediction is feasible for diseases that require daily imaging; for more systemic therapies, that level of monitoring is far too burdensome for a sick patient. However, he stressed that actionable predictions that might help a person in the near term can still be made without knowing everything about a system. Chung posed a question about the ability to bootstrap limited data sets to build out missing time scales. Glazier responded that in many cases, noninvasive measurements are impossible. Therefore, he emphasized that mixed modeling methods are essential—for example, single-cell models of the origin of resistance and continual models of tumor margins. He also proposed modeling organs-on-a-chip systems to be able to measure in better detail, develop confidence in workflows, and understand failure modes. Because mechanistic simulations can generate what is missing from real-world training data, he asserted that combining mechanistic and data-driven approaches could lead to the development of more reliable models.

Chung inquired about the different challenges that arise when using DTs for diagnosis, prognosis, therapy optimization, and drug discovery. Reflecting on this hierarchy of difficulty, Glazier observed that drug discovery is challenging owing to a lack of treatment models. Furthermore, models act at the molecular level even though the responses of interest occur at the system level, and many levels of interaction exist between the two. Measuring and controlling at the same level of the outcome of interest is impossible in medicine; he noted

APPENDIX D

that bulk aggregation could be a useful approach except when two patients who look identical to the model behave differently.

Chung asked about emerging technologies that could be leveraged for DTs as well as significant research gaps that should be addressed. Koumoutsakos added that data obtained from medical applications are often heterogeneous, and a difficult question remains about how to combine them to create DTs. He also emphasized the need to create a common language between the people who take the measurements and those who do the modeling.

DIVERSITY OF MODELS AND DATA IN DIGITAL TWINS

During the third panel, workshop participants heard brief presentations on DTs at the whole-human, multisystem, and population scale from the perspectives of innovative regulatory evaluation and methods development. Much of the conversation centered around the diversity of models and data in DTs.

Aldo Badano, Food and Drug Administration (FDA), indicated that FDA has developed tools to de-risk medical device development and facilitate the evaluation of new technology using least burdensome methods. FDA's Catalog of Regulatory Science Tools[5] includes the Virtual Family[6] and the open-source computational pipeline from the first all-in silico trial for breast imaging.

Badano explained that manufacturers are expected to establish the safety and effectiveness of their devices for intended populations—usually via clinical trials. In a digital world, this requires sampling large cohorts of digital humans to enroll digital populations, which can be data-driven and/or knowledge-based. He described knowledge-based models as the most promising for device assessment because they can be incorporated into imaging pipelines that contain device models. He emphasized that digital families have been used for safety determinations in more than 500 FDA submissions, and

[5] The website for FDA's Catalog of Regulatory Science Tools is https://www.fda.gov/medical-devices/science-and-research-medical-devices/catalog-regulatory-science-tools-help-assess-new-medical-devices, accessed February 27, 2023.
[6] The Virtual Family is a package of extremely detailed, anatomically accurate, full-body computer models used in simulations for medical device safety.

digital cohorts could be used for in silico device trials. Badano postulated that properly anonymized DT data sets could become rich sources of data for generating digital humans in the future. He added that if DTs are eventually embedded into devices for personalized medicine, the DT model itself would likely need to be incorporated into the regulatory evaluation.

David Miller, Unlearn.AI, commented that despite the skepticism surrounding the use of DTs, they have much scientific potential, especially for the development of prognostic DTs. Prognostic DTs provide a rich set of explanatory data for every participant in a randomized clinical trial: each individual in the trial receives a rich set of predictions (i.e., a multivariate distribution) for all outcomes of interest. He stressed that combining prognostic DTs and real participants enables faster and smaller trials.

Miller explained that this method of applying an ML model to historical data and dropping its predictions into a clinical trial without adding bias (known as PROCOVA™) has been qualified by the European Medicines Agency as suitable for primary analysis of Phase 3 pivotal studies. The procedure has three steps: (1) training and evaluating a prognostic model to predict control outcomes, (2) accounting for the prognostic model while estimating the sample size required for a prospective study, and (3) estimating the treatment effect from a completed study using a linear model while adjusting for the control outcomes predicted by the prognostic model. He emphasized that the context of use for DTs is essential: with zero-trust AI, high correlations provide the greatest value, but weak correlations cause no harm; with minimal-trust AI, low bias (for which a confirming mechanism exists) provides the greatest value; and with high-trust AI, precision medicine could replace clinical judgment.

Discussion

Incorporating questions from workshop participants, Thiele moderated a discussion among both speakers and two experts in bioengineering and personalized oncology, respectively: Todd Coleman, Stanford University, and Heiko Enderling, H. Lee Moffitt Cancer Center. Thiele inquired how simulations, computational models, and DTs should be distinguished from one another. Enderling highlighted an overlap in the definitions: DTs need simulation, mechanistic modeling, AI, and ML to answer different types of questions at different scales. He emphasized that synergizing approaches leads to successful outcomes; for example, AI could be used to make mechanistic models more predictive. Coleman suggested that the research community apply its knowledge of anatomy and physiology to develop better generative statistical models and a better understanding of sources of variability. Badano noted that DTs, in silico methods, and AI are distinct concepts that should not be described interchangeably.

Thiele asked about the roles of real-time data, continuous updates, and recalibration for DTs. Coleman stressed that context matters: for a clinical trial related to a neuropsychiatric condition, the DT should be able to predict abrupt changes in behavior and physiology and use that information. He added that the need for continuous updates depends on the richness of the statistics of the population of interest. For applications like PROCOVA™, Miller noted that real-time feedback is not ideal. Enderling and Coleman pointed out that the definition of "real time" (in terms of data collection) is based on the disease of interest. Enderling commented that when studying hormone therapy for prostate cancer, collecting prostate-specific antigen measurements every 2–3 weeks means that the model can be run in between collections, which is sufficient to guide therapy.

Thiele posed a question about the roles of model and parameter standardization for DTs. Enderling replied that calibration and validation are difficult with limited data; he urged the community to identify common objectives and develop rigorous standards to create more trustworthy models. Miller highlighted the value of out-of-sample evaluations, which should match the target population. Badano commented on the need to balance individual validation efforts with population validation efforts.

In response to a question from Thiele about ownership of patient data, Miller stressed that patients should have more control over how their data are shared in clinical trials and used for DTs. Thiele wondered how patients and clinicians could develop trust in DTs. Enderling posited that ensemble models could be used to quantify uncertainty and predictive power and noted that a DT's predictions should be used as a "communication tool" between the clinician and the patient. Miller championed sharing success stories that start with simple problems: patients and clinicians will not trust DTs until the scientific community has fully endorsed them. Coleman focused on the need to learn from past failures and on ensuring that anatomy and physiology is reflected accurately in algorithms He asserted that in the case of medical applications, knowledge about anatomy and physiology should be reflected accurately in algorithms. Badano explained that simulation enables exploration of patients at the edge of the distribution, which offers another avenue to better understand failure modes. Enderling added that education and transparency within the biomedical community (e.g., developing standardized terminology) are essential before successfully communicating across disciplines and educating clinicians and other stakeholders about the potential of DTs.

Thiele posed a question about the challenges of estimating uncertainties from the population level to the individual level. Badano described a benefit of in silico methods: once the models are running, more data can be integrated, enabling the study of a narrower group without increasing the duration of clinical trials. Coleman referenced opportunities for optimal experimental design and sequential experimental design to improve the performance of digital systems at both the population and individual levels. Miller added that prediction methods at the individual level are in a different regulatory category than those for the population level, which could present barriers to innovation.

Thiele highlighted the interdisciplinarity of DTs and wondered about the potential to develop a culture in which publishing failures is acceptable. Miller doubted that financial incentives would emerge to encourage the private sector to publish failures, but he urged peer reviewers in academia to publish papers with negative findings because they could improve models and advance science. Badano suggested modifying the focus from "failures" to "constraints on the results of a model." Enderling proposed creating a database of negative clinical trials with explanations to help build better predictive models.

CONNECTING DATA AND MODELS ACROSS SCALES

During the fourth panel, workshop participants heard brief presentations on the challenges and opportunities associated with leveraging data across scales and multiscale modeling.

Bissan Al-Lazikani, MD Anderson Cancer Center, defined DTs as computational models that support a clinician's decision-making about a specific patient at a specific time. The near-term goal for DTs is the development of fit-for-purpose models, with ambitions for the use of multiscale, comprehensive DTs in the future. Current challenges include questions about how to connect DTs across scales, to address vast missing data, and to account for confidence and error.

Liesbet Geris, University of Liège, described the use of intracellular modeling to understand how to shift osteoarthritis patients from a diseased state to a healthy state—a combination of data-driven and mechanistic models enabled a computational framework that could be validated with in vitro experiments. She explained that osteoarthritis is a disease not only of the cells but also of the tissues and the skeleton. A multiscale model of osteoarthritis allows information to be passed from the tissue to the intracellular level, helping to determine whether a cell will stay healthy or remain diseased, and validation occurs at every level of interaction.

Geris also provided an overview of the EDITH[7] project, which is building an ecosystem for DTs in health care. The project's 2022-2024 goal is to build a vision for and roadmap to an integrated multiscale, -time, and -discipline digital representation of the whole body that could eventually enable the comprehensive characterization of the physiological and pathological state in its full heterogeneity and allow patient-specific predictions for the prevention, prediction, screening,

diagnosis, and treatment of a disease as well as the evaluation, optimization, selection, and personalization of intervention options. Other project goals include identifying research challenges and enabling infrastructure, developing use cases, and driving large-scale adoption, all of which require advances in technology; user experience; ethical, legal, social, and regulatory frameworks; and sustainability.

Gary An, University of Vermont, noted that some medical DT tasks related to prognosis, diagnosis, and optimization of existing drugs are "scale agnostic." However, in other cases, such as drug development, understanding how systems cross scales is important. A bottleneck arises in drug development when trying to determine whether the drug that is expected to work will actually work in patients. To reduce the occurrence of failed clinical trials, he suggested enhancing the evaluation process. To evaluate the potential effect of manipulating a particular pathway (proposed for a novel clinical context) as it manifests at a patient level, he asserted that a mechanistic, multiscale model may be needed.

An emphasized that cell-mechanism, multiscale DTs present several challenges, and many questions remain about epistemic uncertainty/incompleteness, high-dimensional parameter spaces, model identifiability/uncertainty, composability/modularity, and useful intermediate applications. He shared a demonstration of a cell-mechanism, multiscale model of acute respiratory distress syndrome that could become a DT if data interfaces to the real world could be established.

Al-Lazikani elaborated that bottlenecks in drug discovery arise owing to the challenges of multidisciplinary and multiscale data integration and multiparameter optimization. To alleviate the issues associated with integrating data from disparate disciplines that span scales, instead of integrating the data points themselves, she suggested integrating *how* all of the data points interact with each other—essentially establishing edges that can be modeled graphically.[8] This approach, which is especially useful

---

[7] The EDITH website is https://www.edith-csa.eu, accessed February 27, 2023.

[8] Examples of this approach can be viewed on the canSAR.ai website at https://cansar.ai, accessed February 27, 2023.

when data are sparse, is advantageous in that different data are captured in the same logic. It is particularly promising for identifying drug-repurposing opportunities and novel therapeutics for cancers such as uveal melanoma. However, a disadvantage emerges with the need to estimate the impact of what is missing; in this case, she asserted that a hybrid approach of data-driven and mechanism-based models would be useful.

Discussion

Incorporating questions from workshop participants, Rebecca Willett, University of Chicago, moderated a discussion with the panelists. In response to a question from Willett about the technical challenges of building multiscale models, Geris pointed out that challenges at the single scale persist when connecting across scales, and finding a way for the scales to communicate is difficult. She stressed that verification, validation, and uncertainty quantification are critical both at the single scale and for the couplings across scales. The EDITH platform aims to establish agreements on the information that should be passed from one scale (or one organ) to another as well as a common language that can be used across modeling platforms and scales. Willett asked about strategies to ensure that a molecular- or cellular-scale model could be plugged in to a multiscale framework with multiple objectives. Al-Lazikani proposed standardizing the interoperability layers instead of standardizing the data or the models themselves. An observed that such issues of composability are an active research gap.

Reflecting on the use of mechanistic models for drug development to target a particular pathway, Willet posed a question about how to build a DT when the pathway is unknown as well as about the value-added of a DT when the pathway is known. An replied that mechanistic computational models are used with perpetually imperfect knowledge to make predictions at a higher level. Because it will never be possible to know everything, he suggested compartmentalizing what is known and validating that knowledge at both a system level and a clinically relevant population level that captures heterogeneity across the population— which is essential for simulating the intervention side of a prospective clinical trial. Al-Lazikani explained that although pathways were useful for drug discovery 20 years ago, now they are insufficient for treating incredibly rare, heterogeneous diseases. A DT of a subpopulation with a rare disease carries the clinical population throughout, so clinical trials can essentially be done in silico to better determine the potential effectiveness of a drug. An described this as an iterative process, with insight into latent functions through the response to perturbations in the clinical setting.

Willett inquired about the role of causal relationships when building models. Geris replied that, in the study of osteoarthritis, information about joint mechanics will interact with the cellular level and the major pathways; thus, both the engineering and systems biology perspectives should work in tandem.

IMPLEMENTATION ISSUES SURROUNDING DATA PRIVACY IN DIGITAL TWINS

Incorporating questions from workshop participants, Qualters moderated the workshop's final panel discussion among three experts on privacy and ethical considerations for collecting and using data for biomedical DTs: Jodyn Platt, University of Michigan; Lara Mangravite, HI-Bio; and Nathan Price, Thorne HealthTech.

Qualters asked about the benefits of aggregating data for medical DTs and whether additional privacy issues arise beyond those normally associated with health data. Platt highlighted the need to maintain flexibility between knowing what the right data are and having an adaptable data ecosystem, and navigating issues of privacy—protecting individuals from harm should be the primary focus. Mangravite championed the benefits of engaging patients and their care teams more proactively in a conversation about managing care with DTs. She added that collecting data in real time for decision-making creates data governance issues. Price emphasized that combining data threads for DTs could drastically improve personalized medicine; however, increased knowledge about individuals has implications for privacy. He observed that even now, one can push a button on a digital wearable to allow or prevent data

APPENDIX D

access across many devices. He reflected on the value of data governance and stressed that all uses of data for DTs should be made known to, understood by, and controlled by the affected individual.

Qualters pointed out that when data are being collected in real time to improve a model and its predictability, bias could creep into those models if significant populations are excluded. Platt urged attention toward issues of equity that arise in terms of individuals' access to the health care system and to medical technology itself. Patient and community engagement helps to build trust in the benefits of new technology, strengthen accountability, and empower individuals.

Qualters wondered how model uncertainty translates into patient guidance. Price responded that although medical DTs are not yet deployed often in the real world, guidance is based on the data collected at the molecular scale, connected with patient outcome data. Although today this process is human-centric, in the future it might be controlled primarily by algorithms. Therefore, he continued, a relationship with the patient that continues over time is essential to understand prediction accuracy, and guiding the DT feedback loop could help safeguard patients. Qualters asked about non-patient stakeholders (e.g., clinicians, health systems, regulators) who will also have to deal with the uncertainty of DT predictions. Platt encouraged the health system to prepare for the DT experience, as patients expect their doctors to understand and communicate results from the digital space. Mangravite agreed that the opportunity to use DTs for decision-making affects the entire health care system. She stressed that DTs are currently designed to be decision-support tools for humans, who can evaluate and act on the uncertainties.

Qualters invited the panelists to identify research gaps related to privacy and ethical concerns for DTs. Price mentioned the issue of identifiability and supported a focus on the uses of the data. Data-sharing mechanisms are also needed, especially when data are aggregated and models include a person's complete health history. He stressed that an individual's DT will never be completely de-identifiable. Mangravite observed research gaps related to data aggregation and interoperability. Once a governance model is developed to control data use, questions will arise about appropriate data and model access.

Qualters inquired about insurability and other economic impacts related to DT predictions. According to Platt, some people fear that their cost of health care will increase or that they will be less insurable based on the DT's results. Price agreed that insurance companies could benefit by leveraging DTs, and appropriate regulations would be needed. However, he added that if DTs advance preventive medicine, insurance companies that do not adopt them will lose members. He urged workshop participants to focus not only on the challenges of medical DTs but also on the exciting opportunities, such as helping to prevent serious illness in the future.

DISCLAIMER This Proceedings of a Workshop – in Brief was prepared by **Linda Casola** as a factual summary of what occurred at the workshop. The statements made are those of the rapporteur or individual workshop participants and do not necessarily represent the views of all workshop participants; the planning committee; or the National Academies of Sciences, Engineering, and Medicine.

COMMITTEE ON FOUNDATIONAL RESEARCH GAPS AND FUTURE DIRECTIONS FOR DIGITAL TWINS Karen E. Willcox (*Chair*), Oden Institute for Computational Engineering and Sciences, The University of Texas at Austin; **Derek Bingham**, Simon Fraser University; *Caroline Chung*, MD Anderson Cancer Center; **Julianne Chung**, Emory University; **Carolina Cruz-Neira**, University of Central Florida; **Conrad J. Grant**, Johns Hopkins University Applied Physics Laboratory; **James L. Kinter III**, George Mason University; **Ruby Leung**, Pacific Northwest National Laboratory; **Parviz Moin**, Stanford University; **Lucila Ohno-Machado**, Yale University; **Colin J. Parris**, General Electric; *Irene Qualters*, Los Alamos National Laboratory; **Ines Thiele**, National University of Ireland, Galway; **Conrad Tucker**, Carnegie Mellon University; *Rebecca Willett*, University of Chicago; and **Xinyue Ye**, Texas A&M University-College Station. ** Italic indicates workshop planning committee member.*

REVIEWERS To ensure that it meets institutional standards for quality and objectivity, this Proceedings of a Workshop – in Brief was reviewed by **Ahmet Erdemir**, Computational Biomodeling Core, Cleveland Clinic; **Irene Qualters**, Los Alamos National Laboratory; and **Adrian Wolfberg**, National Academies of Sciences, Engineering, and Medicine. **Katiria Ortiz**, National Academies of Sciences, Engineering, and Medicine, served as the review coordinator.

STAFF Samantha Koretsky, Workshop Director, Board on Mathematical Sciences and Analytics (BMSA); **Brittany Segundo**, Study Director, BMSA; **Kavita Berger**, Director, Board on Life Sciences; **Beth Cady**, Senior Program Officer, National Academy of Engineering; **Jon Eisenberg**, Director, Computer Science and Telecommunications Board (CSTB); **Tho Nguyen**, Senior Program Officer, CSTB; **Patricia Razafindrambinina**, Associate Program Officer, Board on Atmospheric Sciences and Climate; **Michelle Schwalbe**, Director, National Materials and Manufacturing Board (NMMB) and Board on Mathematical Sciences and Analytics; and **Erik B. Svedberg**, Scholar, NMMB.

SPONSORS This project was supported by Contract FA9550-22-1-0535 with the Department of Defense (Air Force Office of Scientific Research and Defense Advanced Research Projects Agency), Award Number DE-SC0022878 with the Department of Energy, Award HHSN263201800029I with the National Institutes of Health, and Award AWD-001543 with the National Science Foundation.

This material is based on work supported by the U.S. Department of Energy, Office of Science, Office of Advanced Scientific Computing Research, and Office of Biological and Environmental Research.

This project has been funded in part with federal funds from the National Cancer Institute, National Institute of Biomedical Imaging and Bioengineering, National Library of Medicine, and Office of Data Science Strategy from the National Institutes of Health, Department of Health and Human Services.

Any opinions, findings, conclusions, or recommendations expressed do not necessarily reflect the views of the National Science Foundation.

This proceedings was prepared as an account of work sponsored by an agency of the United States Government. Neither the United States Government nor any agency thereof, nor any of their employees, makes any warranty, express or implied, or assumes any legal liability or responsibility for the accuracy, completeness, or usefulness of any information, apparatus, product, or process disclosed, or represents that its use would not infringe privately owned rights. Reference herein to any specific commercial product, process, or service by trade name, trademark, manufacturer, or otherwise does not necessarily constitute or imply its endorsement, recommendation, or favoring by the United States Government or any agency thereof. The views and opinions of authors expressed herein do not necessarily state or reflect those of the United States Government or any agency thereof.

SUGGESTED CITATION National Academies of Sciences, Engineering, and Medicine. 2023. *Opportunities and Challenges for Digital Twins in Biomedical Research: Proceedings of a Workshop—in Brief*. Washington, DC: The National Academies Press. https://doi.org/10.17226/26922.

Division on Engineering and Physical Sciences

Copyright 2023 by the National Academy of Sciences. All rights reserved.

NATIONAL ACADEMIES Sciences Engineering Medicine

The National Academies provide independent, trustworthy advice that advances solutions to society's most complex challenges.
www.nationalacademies.org

E

Opportunities and Challenges for Digital Twins in Engineering: Proceedings of a Workshop—in Brief

Opportunities and Challenges for Digital Twins in Engineering: Proceedings of a Workshop—in Brief (National Academies of Sciences, Engineering, and Medicine, The National Academies Press, Washington, DC, 2023) is reprinted here in its entirety. The original Proceedings of a Workshop—in Brief is available at https://doi.org/10.17226/26927.

Proceedings of a Workshop—in Brief

Opportunities and Challenges for Digital Twins in Engineering

Proceedings of a Workshop—in Brief

The digital twin is an emerging technology that builds on the convergence of computer science, mathematics, engineering, and the life sciences. Digital twins have potential across engineering domains, from aeronautics to renewable energy. On February 7 and 9, 2023, the National Academies of Sciences, Engineering, and Medicine hosted a public, virtual workshop to discuss characterizations of digital twins within the context of engineering and to identify methods for their development and use. Panelists highlighted key technical challenges and opportunities across use cases, as well as areas ripe for research and development (R&D) and investment. The third in a three-part series, this evidence-gathering workshop will inform a National Academies consensus study on research gaps and future directions to advance the mathematical, statistical, and computational foundations of digital twins in applications across science, medicine, engineering, and society.[1]

PLENARY SESSION 1: DIGITAL TWINS IN STRUCTURAL ENGINEERING

Charles Farrar, Los Alamos National Laboratory (LANL), explained that many digital twins are computer-based digital models of physical systems that interface with data. A ban on system-level nuclear testing[2] as well as increased investments in high-performance computing hardware, code development, and verification and validation methods and experiments enabled initial advances in "digital twin technology" at LANL beginning in 1992. He emphasized that a digital twin is shaped by questions; as those questions evolve, the digital twin evolves to incorporate more detailed physical phenomena, geometry, and data and to account for more sources of uncertainty.

Farrar underscored that validation data are often difficult and costly to obtain and replicating actual loading environments is challenging. All real-world structures have variable properties, and incorporating this variability into modeling is particularly difficult. Most structural models are deterministic but their inputs are often probabilistic. Therefore, he said, uncertainty could be incorporated by varying model parameters based on known or assumed probability distributions.

Farrar indicated that digital twins could include physics-based (e.g., finite element), data-driven (e.g., machine

---

[1] To learn more about the study and to watch videos of the workshop presentations, see https://www.nationalacademies.org/our-work/foundational-research-gaps-and-future-directions-for-digital-twins, accessed February 23, 2023.

[2] In September 1992, the Senate passed the Hatfield-Exon-Mitchell Amendment, a 9-month moratorium on nuclear testing that preceded the Comprehensive Nuclear-Test-Ban Treaty of 1996.

APPENDIX E

learning [ML] and artificial intelligence [AI]), statistical, or hybrid (e.g., physics-constrained ML) models. Structural models are often developed based on nominal geometry and material properties; obtaining data that enable modeling of residual stresses, initial flaws and imperfections, thermal distributions, geometric variability, and details of joints and interfaces is difficult. He remarked on the need to consider time and length scales as well.

Farrar stressed that understanding the limitations of digital twins is critical to success. All models are limited by assumptions associated with the physics being modeled, the training data, the validation data, the knowledge of the physical system, and the knowledge of the inputs to the physical system. These limitations define the domain in which one has confidence in the "answer" provided by the digital twin, although that confidence will not necessarily be uniform across the domain. He described several research gaps and areas for investment, including the following: quantifying the level of model fidelity sufficient to answer questions asked of the digital twin, quantifying the physical system's initial or current conditions and incorporating that information into the digital twin, obtaining data for model validation and uncertainty quantification, developing new approaches to human-computer interfaces, and enhancing education with an interdisciplinary class that focuses on digital twins and emphasizes verification and validation.

Derek Bingham, Simon Fraser University, asked Farrar about the difference between complex multiphysics simulations and digital twins. Farrar explained that digital twins have a tighter integration between the data and the model than simulations.

Bingham also asked about the advantages and disadvantages of data-driven and physics-based models as well as how to better integrate them into digital twins to support decision-making. Farrar replied that modeling operational and environmental changes to systems is difficult; however, data could be acquired with in situ monitoring systems. For physical phenomena that lack first principles models but have ample data, he suggested leveraging data-driven approaches either exclusively or to augment the physical model.

PANEL 1: DIGITAL TWIN USE CASES ACROSS INDUSTRIES

Workshop participants heard brief presentations from and discussions among five panelists, each of whom addressed current methods and practices, key technical challenges and opportunities, and R&D and investment needs related to digital twin use cases across their respective industries. Elizabeth Baron, Unity Technologies, presented perspectives from the automotive industry; Karthik Duraisamy, University of Michigan, focused on computational science and fluid dynamics applications; Michael Grieves, Digital Twin Institute, described the use of digital twins in manufacturing settings; Michael Gahn, Rolls-Royce, discussed aircraft engine design and model-based systems engineering; and Dinakar Deshmukh, General Electric, offered perspectives from the aviation industry.

Current Methods and Practices

Baron noted that the automotive industry has been adapting digital twin technologies since the 1990s and emphasized that increased adoption of real-time digital twins could accelerate Industry 4.0[3] and improve customer-oriented manufacturing, design, and engineering. She defined a digital twin as a dynamic virtual copy of a physical asset, process, system, or environment that behaves identically to its real-world counterpart. It ingests data and replicates processes to predict possible real-world performance outcomes. Processes, tools, and culture are affected, and people play a critical role in testing usability and function in digital twin design. She indicated that digital twins also provide an effective means of communication to account for how people understand and solve problems.

Duraisamy highlighted efforts to train offline dynamic system models to work online—more closely to real time—to attribute causes for events or anomalies; research is under way to create models that both run and make inferences faster. Although challenges related to identifiability, likelihoods and priors, and model errors remain, he noted that many digital twin applications could leverage simple models, algorithms, and decision processes to improve productivity. He described six "classes" of digital twins: Level 1 digital twins provide information to users; Level 2 digital twins assist

---

[3] Industry 4.0, also referred to as the Fourth Industrial Revolution or 4IR.

operators with decision support; Level 3 digital twins empower managers for high-value decision-making with confidence; Level 4 digital twins assist organizations in planning and decision-making; Level 5 digital twins empower organizations to better communicate, plan, and consume knowledge; and Level 6 digital twins define organizations' decisions and create knowledge.

Grieves observed that digital twins typically have three components: (1) the physical space of the product and the environment; (2) the virtual space of the product and the environment; and (3) the connection between the physical and virtual spaces where data from the physical space populate the virtual space, and information developed there is brought back to the physical space.

Gahn described Rolls-Royce's digital engineering framework, where data flow from physical assets to digital models to continually update them. By mirroring physical engineering processes in a digital realm, customers could improve forecasting, reduce life-cycle cost, increase asset availability, and optimize performance. How a company leverages digital twins depends on the desired outcome—for example, eliminating an engineering activity could enable faster, cheaper progress. He emphasized that cybersecurity is a key consideration for digital twins that directly inform decision-making, and a digital twin of the cyberphysical system is needed to protect against adversarial attacks.

Deshmukh provided an overview of commercial digital twin applications for fleet management. When building a representation of an asset for a digital twin, capturing important sources of manufacturing, operational, and environmental variation is key to understanding how a particular component is behaving in the field. Sophisticated representations that capture these sources of variation component by component and asset by asset illuminate how operations could be differentiated. He underscored that reducing disruption, such as unscheduled engine removal or maintenance for major air carriers, is critical in the aviation industry.

Incorporating questions from workshop participants, Parviz Moin, Stanford University, moderated a discussion among the panelists. In response to a question about the security of centralized digital twins. Gahn replied that it depends both on the asset and the types of data. For example, reliability data are usually collated for a digital twin in a central location, but this creates a target for an adversary. Other applications might leverage the edge instead, which offers more security. He explained that securing the digital twin environment is a significant challenge, with data from the asset, data in transit, data stored between those two points, and data for modeling—if the systems are not secure, one cannot develop confidence in those data or the resulting decisions. Baron added that another level of data security to communicate between multifunctional teams would be useful. Moin wondered how proprietary information is handled and whether digital twins should be open source. Duraisamy noted that open sourcing could be beneficial but difficult for some stakeholders, and standards would be useful. Gahn emphasized the value of reference models, and Baron and Duraisamy suggested that a framework could be open source while instances of the data could be highly protected.

Moin inquired about the current state of the art across digital twins. Duraisamy explained that Level 1 digital twins are successful, and applications for Levels 2 and 3 digital twins exist in industry for simple decision-making where systems can be modeled to a high degree of confidence. Duraisamy added that because products are often designed to operate in well-understood regions, parameterization with real-world data could help digital twins become more useful in real-world settings. However, models often are not configured to handle rare events, and geometry changes might not be considered. If the goal is to use digital twins for difficult problems and difficult decisions, he continued, understanding model form error and developing strategies to quantify that error quickly are critical. Deshmukh noted that digital twins should learn continuously from field observations, and Grieves agreed that models will improve with more real-world data for validation.

Moin asked how to quantify the relationship between the amount of data collected and the confidence level in a digital twin's prediction, particularly for low-probability, high-risk events. Duraisamy said that all sources of uncertainty have to be factored in, and understanding whether the algorithms have the capabilities to make a decision within the right time frame is also key. Deshmukh added that ground-truth reality is critical to build digital twins and that models improve with more data, but a framework is needed to update the models continually. Grieves suggested developing a risk matrix to help understand the probability and potential impacts of particular events and to help determine which and how much data are needed. Moin inquired about the norms across use cases that allow one to standardize the quality of a digital twin. Grieves explained that the use case determines whether an online connection is needed for real-time data collection or if data should be collected over months or years.

Moin posed a question about the strength of coordination across component models in terms of model interoperability, data flow, matching data, and frequency of data update needs. Grieves and Baron indicated that coordination is poor. Baron elaborated that many of the data come from multiple sources, have different formats, and have different levels of tolerance in terms of validation and verification criteria. She described this as a difficult but critical problem to solve for digital twins: unifying the data and presenting them in context would allow for consistency throughout the evaluation.

Technical Challenges and Opportunities

Baron explained that an effective digital twin illuminates where systems relate to one another. Providing contextual relationships between these vertical functions[4] (Figure 1) is both a challenge and an opportunity, as every function adds an important level of completeness to the digital twin. Holistic digital twins could build up and break down experiences, using AI to provide information and incorporating human collaboration when insights reveal problems that should be addressed.

Duraisamy observed that discussions about implementing digital twins tend to focus on the sensors, the data,

---

[4] Vertical functions are tasks with performance attributes in systems engineering and design fundamentals with specific requirements that must be satisfied to provide the desired operational capabilities. Each vertical function must also perform with respect to other defined vertical functions for the system to properly work as a whole. (Definition provided by Elizabeth Baron via email on June 7, 2023.)

the models, the human-computer interfaces, and the actions. However, he said that to move from models to actions, the following "enablers" should be considered: uncertainty propagation, fast inference, model error quantification, identifiability, causality, optimization and control, surrogates and reduced-order models, and multifidelity information.

Grieves remarked that manufacturing operations is rich with opportunities for digital twins, where the goal is to use the fewest resources to predict and address future problems. For example, digital twins could improve quality control by enabling virtual testing of a product and help inform a supply network that exists alongside a supply chain. He stressed that interoperability and cybersecurity are essential to prevent threats to the safety of the manufacturing floor.

Gahn elaborated on Rolls-Royce's digital framework, which contains a market twin, program twin, product twin, component twin, production twin, and digital twin. Even within a single organization, each twin requires data from different business units; crossing multiple organizations is even more difficult, with the consideration for protecting intellectual property. Issues might also arise in regulation and certification, and he pointed out that in addition to technical challenges, digital twins have to overcome programmatic, commercial, and legal barriers to ensure the best outcome for the customer.

Deshmukh emphasized that the digital twin "invention to production journey is a team sport." Challenges include defining the scope (i.e., type of problem and model transparency); migrating the analytics into production (i.e., the outcome determines the use of the edge versus the cloud, and the digital twin has to be capable of identifying analytic degradation and uncertainty with time); and considering data availability in production, a scalable data platform, and the right team balance (i.e., data scientists, software engineers, subject-matter experts, and business owners). Deshmukh remarked that opportunities for digital twins include enhanced asset reliability, planned maintenance, reduced maintenance and inspection burden, and improved efficiency. He

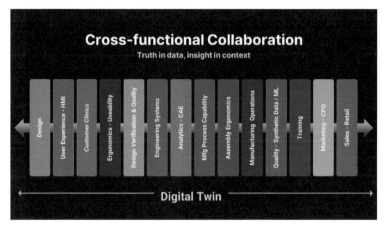

FIGURE 1 Cross-functional collaboration enabled by digital twins. SOURCE: Elizabeth Baron, Unity Technologies, presentation to the workshop, February 7, 2023.

stressed that these outcomes emerge by combining automated, continuous, and curated data; expertise from physics and data science; and capable, scalable, and configurable AI and ML.

Incorporating questions from workshop participants, Carolina Cruz-Neira, University of Central Florida, moderated a discussion among the panelists. She asked about regulatory barriers to digital twin adoption. Grieves explained that risk-averse regulatory agencies need to be educated on the value of digital twins, which could be achieved in part by demonstration. The automotive industry has embraced virtual crash testing because many more types of testing can be done at a much lower cost. Although a physical validation will likely always be required, he asserted that better information emerges via virtual testing (assuming the physics are correct). The most significant barrier to progress in this area, he continued, is the established culture of the test community.

Cruz-Neira posed a question about challenges related to the interoperability of and the mathematical foundations for digital twins. Duraisamy commented that when building a digital twin of a product at different stages of its life cycle, quantities of interest vary. However, the level of information needed and how to transfer that information across parts of the life cycle is not well understood. Mathematical models that provide information beyond these quantities of interest are useful for many applications, he said, including for interoperability. Cruz-Neira inquired about the challenges of sustaining a digital twin over a product's life cycle. Deshmukh said that models might not maintain their level of confidence when put into production; "guardrails" to constantly assess the outcome of the digital twin with respect to the ground-truth verification would be useful. Grieves added that digital twins should be "learning models" that capture information about the degradation of old products to better inform and maintain new products.

In response to a question from Cruz-Neira about the use of digital twins for training, Grieves described this application as a significant opportunity to allow for mistakes to be made in the virtual environment. Gahn said that his team has been pursuing the use of digital twins for training with augmented reality sessions for engine overhaul and maintenance. He added that the

APPENDIX E

ability to arrange the manufacturing floor virtually for equipment placement is another opportunity.

Cruz-Neira asked about nontechnical (i.e., cultural, financial, and managerial) barriers to leveraging the full potential of digital twins. Baron explained that every vertical function (e.g., design, engineering, and manufacturing) has its own culture; if issues emerge where these functions converge, questions arise about responsibility. However, she said that high-level leaders should enable employees to work collectively to find affordable solutions by relying on data for contextual insight. Grieves pointed out that decision-makers are not digital natives, and digital twins "look like magic" to them. He advised educating, training, and building trust among decision-makers, ensuring that they understand the potential of the technology and will invest when appropriate. Duraisamy said that no single algorithm exists to solve all and decision-makers should understand that progress happens gradually by assimilating knowledge and incorporating rigor. Deshmukh suggested connecting business outcomes to technology innovations and encouraging leadership to invest incrementally. Gahn agreed that defining a tangible benefit related to specific stakeholders in a language they understand is critical.

Research and Development and Investment

Baron presented a hierarchy of digital twins that described their potential capabilities. The "virtual twin" is a physically accurate, realistic digital representation of an asset, facility, or product that emulates its real-world counterpart. The "connected twin" integrates real-time and right-time data to provide insights into the performance of an asset at specific points in time, which requires significant human-in-the-loop interaction. The "predictive twin" leverages data to predict outcomes for the operations of complex facilities and equipment. The "prescriptive twin" leverages advanced modeling and real-time simulation for potential future scenarios as well as prescriptive analytics. The "autonomous twin," the "nirvana of digital twins," would use multiple real-time data streams to learn and make decisions to correct issues automatically and enable predictive and prescriptive analytics. She explained that the success of any digital twin relies on people collaborating. The pathway to this success also leverages increased computing capability to solve problems faster, as well as lessons from the past to better predict and address issues in the future.

Duraisamy stated that research on deriving efficient physics-constrained models, as well as those that assimilate information, is needed; ideally, these models would be probabilistic. Efficient and accurate models and algorithms are key to realizing the full potential of digital twins, he added. Investment in methods and algorithms for scalable inference and uncertainty quantification, identifiability, causality, and physics-constrained modeling is also critical—quantifying and effectively managing model form errors and uncertainties remains problematic. He encouraged the community to focus on open standards, common terminology, verification, and validation and championed foundational mathematics to advance digital twins.

Grieves highlighted the need for further research in ontologies and harmonization among groups; interoperability (from cells, to units, to systems, to systems of systems); causality, correlation, and uncertainty quantification; data-physics fusion; and strategies to change the testing and organizational culture. He pointed out that U.S. digital twin research is currently limited as compared to work in Europe and China, and he proposed that digital twins be introduced to students as part of a revised postsecondary curriculum.

Deshmukh explained that data are at the core of digital twin success, and digital twin adoption is the critical end point. He asserted that a scalable platform for structured and unstructured data and the ability to compute at scale are needed. Integrating data science and domain knowledge is critical to enable decision-making based on analytics to drive process change, he said, and tomorrow's "disruptors" will manage massive amounts of data and apply advanced analytics with a new level of intelligent decision-making.

Incorporating questions from workshop participants, Conrad Tucker, Carnegie Mellon University, moderated a discussion among the panelists. He asked them to elaborate on the significance of "interoperability" for digital twins. Duraisamy defined "interoperability" as managing "information" so that it travels seamlessly across the full chain. Baron explained that nothing is developed in a silo: many data have to interoperate in a digital twin for training, and many groups have to work together on a digital twin for problem solving. Grieves noted that machines on the factory floor have an interoperability problem that could be addressed with platforms that "translate" one machine to another. Although standards might not be needed, he continued, harmonization (especially with smaller manufacturers) is essential.

In response to a question from Tucker, Gahn noted that intrusion detection is only one component of protecting digital twins. Many frameworks are available, but a baseline level of protection prepares users with a plan to recover data and modeling tools if an attack occurs. He suggested that users consider protection both in terms of the cloud and the physical asset that moves around globally. A secure download from the asset to the Internet is needed to capture data, he continued, but questions arise about encryption and managing encryption keys.

Tucker asked the panelists how they would convince skeptics to invest sustainably in digital twins. Deshmukh advised focusing on the business problem that could be solved with digital twins: organizations that adopt the technology could offer better experiences for their customers. Gahn added that, to temper expectations, stakeholders should understand what the digital twin will *not* do (e.g., predict something that is not in the model). Grieves observed the difficulty of "selling" future opportunities; instead of presenting technical pitches to stakeholders, he suggested discussing potential financial performance and cost reduction in the near term. Baron proposed presenting information in context— volumetrically, visually, functionally, and personally— within a globally connected team. Duraisamy suggested highlighting the "digital threads" that could make products more versatile.

Tucker posed a question about opportunities for engineering education, and Grieves encouraged an integrative approach: curricula should include a focus on real-world problems, which would increase student understanding, motivation, and retention. Duraisamy pointed out that although the current curriculum, shaped in the 1960s, has value, it lacks several components. Students have become better system-level thinkers, which is an important skill to accompany mathematics and physics knowledge. He emphasized that concepts should be connected across the entire degree program. Gahn explained that Rolls-Royce values "T-shaped" engineers with breadth across disciplines. He suggested new strategies to introduce digital twin concepts, including microtraining or microcertifications. Deshmukh commented that digital upskilling for the current workforce is essential.

PLENARY SESSION 2: DIGITAL TWINS FOR NATIONAL SECURITY AND RENEWABLE ENERGY

Grace Bochenek, University of Central Florida, described digital twins as "innovation enablers" that are redefining engineering processes and multiplying capabilities to drive innovation across industries, businesses, and governments. She asserted that this level of innovation is facilitated by a digital twin's ability to integrate a product's entire life cycle with performance data and to employ a continuous loop of optimization. Ultimately, digital twins could reduce risk, accelerate time from design to production, and improve decision- making as well as connect real-time data with virtual representations for remote monitoring, predictive capabilities, collaboration among stakeholders, and multiple training opportunities.

Bochenek noted that opportunities exist in the national security arena to test, design, and prototype processes and exercise virtual prototypes in military campaigns or with geopolitical analysis to improve mission readiness. Digital twins could increase the speed of delivery for performance advantage, and digital twin technologies could help the United States keep the required pace of innovation by assessing systems against evolving threats and finding solutions. She pointed out that questions remain about how to connect digital twins across an organization to build stakeholder trust and confidence in decisions and investments.

APPENDIX E

Bochenek presented another opportunity to leverage digital twins in the energy sector, which aims to modernize the electric grid with renewable energy against a backdrop of quickly changing regulatory requirements and complex energy systems. She said that the ability to accelerate technology development, to optimize operations, and to use analysis to innovate advanced energy systems at scale is critical. Digital twins could enhance understanding of physical grid assets; help utility companies improve planning; expand training for personnel; and improve the cycle of learning, designing, and testing.

Bochenek summarized that digital twins are decision-making tools that help determine how to use finite resources more efficiently to drive sustainability, develop better ideas, and initiate more productive partnerships with stakeholders. Scale and fidelity are significant digital twin challenges, as is the availability of open, real-time, high-quality data. Data ownership, data management, data interoperability, intellectual property rights, and cybersecurity are also key considerations. She underscored that because energy systems are large and complex, grid modernization demands advanced modeling capabilities and real-time interaction with data for predictive and forensic analysis and energy resource protection.

Incorporating questions from workshop participants, Moin moderated a discussion with Bochenek. He posed a question about the most compelling use cases for digital twins. Bochenek described opportunities to use digital twins to support planning, prediction, and protection for smart cities. Other opportunities exist in homeland security and transportation, but each requires involvement from both local and state governments. Space operations is another exciting application for digital twins, she added.

Moin asked about the progress of implementing digital twin concepts in the postsecondary curriculum. Bochenek responded that developing a workforce that is proficient in and can innovate with digital twin technologies is key to drive the future forward. Because digital twins are interdisciplinary by nature, she stressed that education should be interdisciplinary. Additional thought could be given to the role that humans play in digital twins as well as how to ensure that foundational science informs engineering applications. She also advocated for increased certificate programs and professional education for the current workforce.

PANEL 2: DIGITAL TWIN USE CASES ACROSS INDUSTRIES

Workshop participants heard brief presentations from and discussions among four panelists, each of whom addressed current methods and practices, key technical challenges and opportunities, and R&D and investment needs related to digital twin use cases across their respective industries. José Celaya, SLB (previously Schlumberger), discussed modeling challenges for digital twin value creation in the oil and gas industry; Pamela Kobryn, Air Force Research Laboratory (AFRL), shared her views on aircraft sustainment; Devin Francom, LANL, offered perspectives on stockpile stewardship; and Devin Harris, University of Virginia, discussed large-scale infrastructure systems.

Current Methods and Practices

Celaya pointed out that using the term "digital twin" to describe ongoing computational modeling would not unlock new value or performance. New attributes are critical, and a digital representation could focus on a different perspective of an asset or machine such as reliability. He described digital twins as "living systems" and explained that the large critical assets and the asset fleets that comprise the oil and gas industry require different types of digital twins for optimization. He suggested increased partnership with technology providers to advance data management, use of the cloud, and modeling capabilities for digital twins, which could improve internal productivity and efficiency and enhance products for customers.

Kobryn provided an overview of the AFRL Airframe digital twin program, which focused on better maintaining the structural integrity of military aircraft. The initial goal of the program was to use digital twins to balance the need to avoid the unacceptable risk of catastrophic failure with the need to reduce the amount of downtime for maintenance and prevent complicated and expensive repairs. The program focused on how operators use the fleet—for example, anticipating how they would fly the aircraft and mining past usage data to update the forecasts for each aircraft. Kobryn noted that understanding the "state" of the aircraft was also critical—for instance, any deviations from the manufacturing blueprint, as well as inspections and repairs, would be tracked over the aircraft's lifetime. She explained that all of these data informed simulations to provide timely and actionable information to operators about what maintenance to perform and when. Operators could then plan for downtime, and maintainers could prepare to execute maintenance packages tailored for each physical twin and use them for updates. Moving forward, connecting the simulations across length scales and physical phenomena is key, as is integrating probabilistic analysis. Although much progress has been made, she noted that significant gaps remain before the Airframe digital twin can be adopted by the Department of Defense.

Francom explained that science-based stockpile stewardship emerged after the Comprehensive Nuclear-Test-Ban Treaty of 1996.[5] DOE invested in simulation and experimental capabilities to achieve this new level of scientific understanding that leveraged available data and simulation capacity to ensure the safety and functionality of the stockpile. He noted that uncertainty quantification, which is critical in this approach but difficult owing to the different categories of physics and engineering involved, could be better understood with the use of digital families. For example, digital twins have the ability to leverage global and local data to refine the understanding of relevant physics and enhance the accuracy of predictive simulations. He indicated that many tools could be useful in collaboration with digital twins—for instance, design of experiments, surrogate modeling, sensitivity analysis, dimension reduction, calibration and inversion, multifidelity uncertainty quantification, and prediction and decision theory—but gaps in the state of the art for reasoning about uncertainty could be pursued further.

Harris remarked that smart cities are the model for sociotechnical systems of the future. He defined the use of digital twins for smart cities as virtual representations of engineered systems that allow engineers and operators to understand how a system performs over time and under future scenarios—that is, a decision-making tool with a real-time, cyber-human collaboration framework. He discussed the potential use of digital twins to monitor bridge structures, for which timescales are particularly challenging. In the past, sensing tools were installed to find damage over time, but it was difficult to understand localized damage. Structural health monitoring then shifted from purely depending on "experimental" measures to relying on "models," which are more cost-effective than expensive sensor networks. However, he stressed that tracking aging effects in an operational environment over time is still difficult, and a mechanism that represents the physical system is needed.

Incorporating questions from workshop participants, Cruz-Neira moderated a discussion among the panelists. She posed a question about the difference between a digital twin and a simulation. Celaya replied that a digital twin is a living model that reflects the reality of the physical asset through time. Francom wondered if any value exists in making a concrete distinction between the two; instead, he said that the community should strive for inclusivity, focusing on how the modeling and simulation world could improve digital twins and vice versa. Kobryn agreed and added that the distinction only matters when one wants to spotlight special use cases for digital twins. She pointed out that the aspect of "updating" the digital twin with real-world data makes it unique. Cruz-Neira asked if a point exists when so many data are available on a physical asset that the physics-based model would be abandoned for an empirical model. Harris responded that, historically, that amount of data has never been available for large infrastructure. He explained that a realistic strategy to monitor structural health is to determine the current baseline with existing structures and use temporal measurements over time to understand boundary conditions and loading conditions. Celaya highlighted the value of leveraging physics with statistical inference rather than dynamic systems for these cases.

Cruz-Neira inquired about the critical elements for digital twins in engineering. Kobryn explained that digital twins help make the shift from steady-state,

---

[5] For more information about the treaty, see CTBTO Preparatory Commission, "The Comprehensive Nuclear-Test-Ban Treaty," https://www.ctbto.org/our-mission/the-treaty, accessed April 7, 2023.

APPENDIX E

single-discipline engineering models and analyses to multidisciplinary efforts that blend the best models and simulations from computational and theoretical spaces with real-world data. Harris emphasized the need to consider what these digital twins will look like when integrated into larger systems.

Cruz-Neira posed a question about how to integrate unit-level models with a system-level twin. Francom pointed out that the system level often reveals one aspect of information and the unit level another. When this happens, the models are mis-specified and model form error exists. Because model form error is difficult to track and quantify, he suggested allowing for disagreement in the models and embracing the uncertainty instead of forcing the models to agree, which creates statistical issues.

Cruz-Neira asked how to collaborate with decision-makers to define the requirements of digital twins to ensure that they are helpful. Kobryn asserted that if the end user is not involved with the digital twin design, resources are likely being spent to solve the wrong problem. At AFRL, design thinking and human-centered design are prioritized. Cruz-Neira highlighted the value of making any assumptions transparent to the end user. Francom noted that a spectrum of digital twins would be useful to explore different assumptions about and explanations for certain phenomena. Kobryn added that model-based systems engineering could make these assumptions explicit for an audience with expertise.

Technical Challenges and Opportunities

Celaya explained that challenges and opportunities for digital twins are business case and context dependent. For example, having a modeler in the loop is not scalable for many use cases, and opportunities exist to better leverage computational power and data access. As a result, the data could move from the physical asset to the digital twin in a computational step, and the digital twin would observe how the physical asset is behaving and adapt. He also suggested collecting data with the specific purpose of the digital twin in mind, and he noted that standards and guidelines could help better link the use case to the model requirements. Instead of leveraging old workflows to sustain digital twins, he proposed developing a new paradigm for optimization to unlock the value proposition of Industry 4.0.

Kobryn presented the following four categories of challenges and opportunities for digital twins: (1) working with end users to understand what information they want from a digital twin and how often that information should be updated; (2) determining how to protect personal privacy and intellectual property, how to secure data and models to prevent unauthorized access and maintain prediction validity, and how to address liability for operational failures; (3) determining the level of simulation fidelity and the level of tailoring for individual assets or operators, and strategies to verify and validate probabilistic simulations; and (4) deciding how to reduce computation time and cost and collect an asset's state and usage data reliably and affordably.

Francom highlighted uncertainty quantification as both a challenge and an opportunity for any problem with high-consequence decisions. Extrapolation is particularly difficult when a digital twin in one regime is used to make predictions in a different regime, he noted, and balancing the physics and the empirical components of a digital twin is critical. Quantifying model form error is also a key challenge, especially when one cannot rely solely on empirical information, which is sometimes incorrect. To begin to address these issues, he mentioned that LANL is exploring opportunities to use information from neighboring systems and to cut the feedback loop when information could corrupt a digital twin.

Harris indicated that no good mechanism is available to deal with existing assets, even though most community stakeholders are concerned about maintaining the function and safety of existing infrastructure. Identifying a way to translate these old systems into more modern formats is needed, he said, although challenges will arise with scale. Most government agencies manage their own systems with limited resources, and working with end users to understand the problem the digital twins are being designed to address is critical. He explained that accessing data for critical infrastructure is especially difficult, given that the information is not publicly available for safety reasons, as is creating digital representations based on decades-old structural plans; however, opportunities exist to leverage new forms of data.

Incorporating questions from workshop participants, Tucker moderated a discussion among the panelists. He asked how a decision would be made to cut the feedback loop of a digital twin when the potential for corrupt information exists, and how one would verify that this was the right decision. Kobryn championed active assessment of data quality with automated technology. She cautioned that as a digital twin continues to be updated in real time, more computational challenges will arise, but implementation of zero trust networks could be beneficial. Celaya explained that domain knowledge could be used to validate information and check for corrupted sensors, corrupted data transportation, or noise. He emphasized the value of both technology and science to address this issue. Tucker wondered how a practitioner would determine whether the source of error is the sensor or the digital twin. Francom noted that the level of trust in the physics and the data is influenced by the fact that the data collection could be flawed. He added that industrial statistics and control charts could help detect whether the sensors are malfunctioning or the system itself is changing. Harris explained that because systems often work off of several sensors, opportunities exist to cross-reference with other sensors to determine whether the sensor or the model is causing the error. Kobryn suggested leveraging Bayesian statistical techniques, where new data are not overweighted. Celaya proposed that models be enhanced to reflect degradation phenomena of systems over time.

Tucker posed a question about the value of optimal experimental design, active learning, optimal sensor placement, and dynamic sensor scheduling. Kobryn described these as significant areas of opportunity for digital twins. For example, by using simulations to determine which test conditions to run and where to place sensors, physical test programs could be reduced and digital twins better calibrated for operation. Francom observed that there is much to be gained from these types of technologies, especially given the high cost of sensors. He highlighted the need to understand what is learned from a sensor; in some cases, sensor capability could be relaxed, but that kind of exercise requires thinking carefully about uncertainty.

Tucker invited the panelists to summarize the most significant opportunities in the digital twin ecosystem. Kobryn detailed an opportunity for product engineering to use digital twin techniques to develop a modeling certification basis for engineering parameters and to validate and design models for deployment. Francom recognized that researchers can learn much from data to improve modeling capabilities; a tighter integration of data and models could help pursue new scientific endeavors. Harris urged the community to gather demonstration cases and to cooperate with clients to understand the parameters in which they work and how a digital twin could solve their problem. Celaya highlighted opportunities for new types of optimization—in sustainability, decarbonization, and environmental protection—as well as for more frequent optimization with digital twins.

Research and Development and Investment

Celaya noted that significant modeling work remains for science to lead to engineering synthesis. He pointed out that methods to address trade-offs (e.g., model utility versus computational cost and speed versus uncertainty) are needed and expressed his support of software-defined modeling as well as contextualizing strong engineering principles in the current drive for empiricism. He encouraged increased investments to enable the scalability of new research as well as to exploit new computation paradigms (e.g., cloud, edge, and Internet of Things) with distributed models. Reflecting on which R&D areas should *not* receive investments could also be a valuable exercise, he added.

Kobryn described several "enabling technologies" for digital twins. She said that leveraging advances in computer engineering and cloud and mobile computing could help to develop architecture for computing, storing, and transporting; and advances in three-dimensional scanning and sensor technology could help advance understanding of the state of physical systems. The Internet of Things and the democratization of data

APPENDIX E

could also be leveraged, and many opportunities exist in physics-based modeling and simulation to improve sensemaking and interpretation of data. Additional opportunities exist to take advantage of big data and AI, but she cautioned against relying too much on either pure data-driven approaches or first principles physics and instead encouraged fusing the two approaches in a modern architecture.

Francom underscored that investments should be made in uncertainty quantification. He described the need both to consider the information that is entering a digital twin and to recognize the many possible ways to obtain the same answer (i.e., nonidentifiability), which is reflected as uncertainty. He also highlighted opportunities for people with different expertise to collaborate to address the challenges inherent in these nonlinear systems.

Harris encouraged investments to enable smart systems, which should be built and preserved with the future in mind to be sustainable, agile, resilient, and durable. These systems could be realized by leveraging automation, advanced sensing, data analytics, and community-centered engagement. He explained that strategic investment is also needed for successful use cases that demonstrate how end users benefit from digital twins, for corporate partnerships, and for interdisciplinary collaboration. Digital twins could progress with increased investment in the void between basic and applied research, he added.

Incorporating questions from workshop participants, Bingham moderated a discussion among the panelists. He wondered if as digital twins become more accurate, users will confuse the "map" they offer for reality. Francom reflected on the problems that could arise if a decision-maker fails to understand that the digital twin is not reality; this is an example of the value of transparency and communication about digital twins. Kobryn highlighted the need for general training in digital literacy, asserting that many decision-makers have nonscientific expertise. She cautioned against overloading the nontechnical workforce with technical information and proposed helping them understand how to use (or not use) data and develop the critical thinking to understand when information can (and cannot) be trusted.

Bingham inquired about top-priority investment areas. Harris championed large-scale investment in a series of interdisciplinary team efforts and in the human aspect of digital twins. In response to a question from Bingham about investments in high-performance computing and efficient algorithms to deal with the complexity and scale of digital twin applications, Kobryn proposed using the high-fidelity capability from high-performance computing to synthesize data to train reduced-order models to take advantage of available data and computing capability at the operating end. Multidisciplinary teams are needed to reduce the order of models to account for relevant physics while leveraging AI and ML, she said.

Bingham asked if investments in education would be beneficial. Instead of creating an entirely "digital curriculum," Celaya suggested enhancing the current core engineering curriculum with a new focus on computing capability, disparate sources of data, and uncertainty to improve students' data dexterity and analytical skills. Kobryn proposed that educators focus on systems engineering in context and provide more real-world experiences within multidisciplinary teams (e.g., Capstone projects) to better prepare students for the workforce. Francom suggested that educators build these real-world examples from weather and stock market data. Harris said that the Capstone experience is valuable, but more industry-academia collaborations would be beneficial. Celaya described an opportunity to teach students when to employ empirical science to inform decision-making. He reiterated the value of developing systems thinking, accompanied by open-mindedness, to model reality at different stages of abstraction.

DISCLAIMER This Proceedings of a Workshop—in Brief was prepared by **Linda Casola** as a factual summary of what occurred at the workshop. The statements made are those of the rapporteur or individual workshop participants and do not necessarily represent the views of all workshop participants; the planning committee; or the National Academies of Sciences, Engineering, and Medicine.

COMMITTEE ON FOUNDATIONAL RESEARCH GAPS AND FUTURE DIRECTIONS FOR DIGITAL TWINS Karen Willcox (*Chair*), Oden Institute for Computational Engineering and Sciences, The University of Texas at Austin; **Derek Bingham**, Simon Fraser University; **Caroline Chung**, MD Anderson Cancer Center; *Julianne Chung, Emory University*; **Carolina Cruz-Neira**, University of Central Florida; **Conrad J. Grant**, Johns Hopkins University Applied Physics Laboratory; **James L. Kinter III**, George Mason University; **Ruby Leung**, Pacific Northwest National Laboratory; *Parviz Moin, Stanford University*; **Lucila Ohno-Machado**, Yale University; **Colin J. Parris**, General Electric; **Irene Qualters**, Los Alamos National Laboratory; **Ines Thiele**, National University of Ireland, Galway; **Conrad Tucker**, Carnegie Mellon University; **Rebecca Willett**, University of Chicago; and **Xinyue Ye**, Texas A&M University–College Station. * *Italic indicates workshop planning committee member.*

REVIEWERS To ensure that it meets institutional standards for quality and objectivity, this Proceedings of a Workshop—in Brief was reviewed by **Burcu Akinci**, Carnegie Mellon University; **Nia Johnson**, National Academies of Sciences Engineering, and Medicine; and **Parviz Moin**, Stanford University. **Katiria Ortiz**, National Academies of Sciences, Engineering, and Medicine, served as the review coordinator.

STAFF Beth Cady, Senior Program Officer, National Academy of Engineering, and **Tho Nguyen**, Senior Program Officer, Computer Science and Telecommunications Board (CSTB), Workshop Directors; **Brittany Segundo**, Program Officer, Board on Mathematical Sciences and Analytics (BMSA), Study Director; **Kavita Berger**, Director, Board on Life Sciences; **Jon Eisenberg**, Director, CSTB; **Samantha Koretsky**, Research Assistant, BMSA; **Patricia Razafindrambinina**, Associate Program Officer, Board on Atmospheric Sciences and Climate; **Michelle Schwalbe**, Director, National Materials and Manufacturing Board (NMMB) and BMSA; **Erik B. Svedberg**, Senior Program Officer, NMMB; and **Nneka Udeagbala**, Associate Program Officer, CSTB.

SPONSORS This project was supported by Contract FA9550-22-1-0535 with the Department of Defense (Air Force Office of Scientific Research and Defense Advanced Research Projects Agency), Award Number DE-SC0022878 with the Department of Energy, Award HHSN263201800029I with the National Institutes of Health, and Award AWD-001543 with the National Science Foundation.

This material is based on work supported by the U.S. Department of Energy, Office of Science, Office of Advanced Scientific Computing Research, and Office of Biological and Environmental Research.

This project has been funded in part with federal funds from the National Cancer Institute, National Institute of Biomedical Imaging and Bioengineering, National Library of Medicine, and Office of Data Science Strategy from the National Institutes of Health, Department of Health and Human Services.

Any opinions, findings, conclusions, or recommendations expressed do not necessarily reflect the views of the National Science Foundation.

This proceedings was prepared as an account of work sponsored by an agency of the United States Government. Neither the United States Government nor any agency thereof, nor any of their employees, makes any warranty, express or implied, or assumes any legal liability or responsibility for the accuracy, completeness, or usefulness of any information, apparatus, product, or process disclosed, or represents that its use would not infringe privately owned rights. Reference herein to any specific commercial product, process, or service by trade name, trademark, manufacturer, or otherwise does not necessarily constitute or imply its endorsement, recommendation, or favoring by the United States Government or any agency thereof. The views and opinions of authors expressed herein do not necessarily state or reflect those of the United States Government or any agency thereof.

SUGGESTED CITATION National Academies of Sciences, Engineering, and Medicine. 2023. *Opportunities and Challenges for Digital Twins in Engineering: Proceedings of a Workshop—in Brief*. Washington, DC: The National Academies Press. https://doi.org/10.17226/26927.

Division on Engineering and Physical Sciences

Copyright 2023 by the National Academy of Sciences. All rights reserved.

NATIONAL ACADEMIES
Sciences
Engineering
Medicine

The National Academies provide independent, trustworthy advice that advances solutions to society's most complex challenges.

www.nationalacademies.org

F

Committee Member Biographical Information

KAREN E. WILLCOX, *Chair*, is the director of the Oden Institute for Computational Engineering and Sciences, an associate vice president for research, and a professor of aerospace engineering and engineering mechanics at The University of Texas (UT) at Austin. Dr. Willcox is also an external professor at the Santa Fe Institute. At UT, she holds the W.A. "Tex" Moncrief, Jr. Chair in Simulation-Based Engineering and Sciences and the Peter O'Donnell, Jr. Centennial Chair in Computing Systems. Before joining the Oden Institute in 2018, Dr. Willcox spent 17 years as a professor at the Massachusetts Institute of Technology (MIT), where she served as the founding co-director of the MIT Center for Computational Engineering and the associate head of the MIT Department of Aeronautics and Astronautics. Prior to joining the MIT faculty, she worked at Boeing Phantom Works with the Blended-Wing-Body aircraft design group. Dr. Willcox is a fellow of the Society for Industrial and Applied Mathematics and a fellow of the American Institute of Aeronautics and Astronautics, and in 2017 she was appointed a member of the New Zealand Order of Merit for services to aerospace engineering and education. In 2022, she was elected to the National Academy of Engineering. Dr. Willcox is at the forefront of the development and application of computational methods for design, optimization, and control of next-generation engineered systems. Several of her active research projects and collaborations with industry are developing core mathematical and computational capabilities to achieve predictive digital twins at scale.

DEREK BINGHAM is a professor and chair of the Department of Statistics and Actuarial Science at Simon Fraser University. He received his PhD from the Department of Mathematics and Statistics at Simon Fraser University in

1999. After graduation, Dr. Bingham joined the Department of Statistics at the University of Michigan. He moved back to Simon Fraser in 2003 as the Canada Research Chair in Industrial Statistics. He has recently completed a 3-year term as the chair for the Natural Sciences and Engineering Research Council of Canada's Evaluation Group for Mathematical and Statistical Sciences. The focus of Dr. Bingham's current research is developing statistical methods for combining physical observations with large-scale computer simulators. This includes new methodology for Bayesian computer model calibration, emulation, uncertainty quantification, and experimental design. Dr. Bingham's work is motivated by real-world applications. Recent collaborations have been with scientists at U.S. national laboratories (e.g., Los Alamos National Laboratory), Department of Energy–sponsored projects (Center for Exascale Radiation Transport), and Canadian Nuclear Laboratories.

CAROLINE CHUNG is the vice president and chief data officer and is an associate professor in radiation oncology and diagnostic imaging at the MD Anderson Cancer Center. Her clinical practice is focused on central nervous system malignancies, and her computational imaging laboratory has a research focus on quantitative imaging and computational modeling to detect and characterize tumors and toxicities of treatment to enable personalized cancer treatment. Internationally, Dr. Chung is actively involved in multidisciplinary efforts to improve the generation and utilization of high-quality, standardized imaging to facilitate quantitative imaging integration for clinical impact across multiple institutions, including as the vice chair of the Radiological Society of North America Quantitative Imaging Biomarkers Alliance and co-chair of the Quantitative Imaging for Assessment of Response in Oncology Committee of the International Commission on Radiation Units and Measurements. Beyond her clinical, research, and administrative roles, Dr. Chung enjoys serving as an active educator and mentor with a passion to support the growth of diversity, equity, and inclusion in science, technology, engineering, and mathematics, including in her role as the chair of Women in Cancer, a not-for-profit organization that is committed to advancing cancer care by encouraging the growth, leadership, and connectivity of current and future oncologists, trainees, and medical researchers. Her recent publications include work on building digital twins for clinical oncology.

JULIANNE CHUNG is an associate professor in the Department of Mathematics at Emory University. Prior to joining Emory in 2022, Dr. Chung was an associate professor in the Department of Mathematics and part of the Computational Modeling and Data Analytics Program at Virginia Tech. From 2011 to 2012, she was an assistant professor at The University of Texas at Arlington, and from 2009 to 2011 she was a National Science Foundation (NSF) Mathematical Sciences Postdoctoral Research Fellow at the University of Maryland, College Park. She received her PhD in 2009 in the Department of Math and Computer Science at

Emory University, where she was supported by a Department of Energy Computational Science Graduate Fellowship. Dr. Chung has received many prestigious awards, including the Frederick Howes Scholar in Computational Science award, an NSF CAREER award, and an Alexander von Humboldt Research Fellowship. Her research interests include numerical methods and software for computing solutions to large-scale inverse problems, such as those that arise in imaging applications.

CAROLINA CRUZ-NEIRA is a pioneer in the areas of virtual reality and interactive visualization, having created and deployed a variety of technologies that have become standard tools in industry, government, and academia. Dr. Cruz-Neira is known worldwide for being the creator of the CAVE virtual reality system. She has dedicated part of her career to transferring research results into daily use by spearheading several open-source initiatives to disseminate and grow virtual reality technologies and by leading entrepreneurial initiatives to commercialize research results. Dr. Cruz-Neira has more than 100 publications, including scientific articles, book chapters, magazine editorials, and others. She has been awarded more than $75 million in grants, contracts, and donations. She is also recognized for having founded and led very successful virtual reality research centers, including the Virtual Reality Applications Center at Iowa State University, the Louisiana Immersive Technologies Enterprise, and now the Emerging Analytics Center. Dr. Cruz-Neira has been named one of the top innovators in virtual reality and one of the top three greatest women visionaries in this field. *BusinessWeek* magazine identified her as a "rising research star" in the next generation of computer science pioneers; she has been inducted as a member of the National Academy of Engineering, is an Association for Computing Machinery Computer Pioneer, and received the IEEE Virtual Reality Technical Achievement Award and the Distinguished Career Award from the International Digital Media and Arts Society, among other national and international recognitions. Dr. Cruz-Neira has given numerous keynote addresses and has been the guest of several governments to advise on how virtual reality technology can help to give industries a competitive edge leading to regional economic growth. She has appeared on numerous national and international television shows and podcasts as an expert in her discipline, and several documentaries have been produced about her life and career. She has several ongoing collaborations in advisory and consulting capacities on the foundational role of virtual reality technologies with respect to digital twins.

CONRAD J. GRANT is the chief engineer for the Johns Hopkins University Applied Physics Laboratory (APL), the nation's largest University Affiliated Research Center, performing research and development on behalf of the Department of Defense, the intelligence community, the National Aeronautics and Space Administration, and other federal agencies. He previously served for more than a decade as the head of the APL Air and Missile Defense Sector, where he led 1,200 staff developing advanced air and missile defense systems for the U.S. Navy and the Missile Defense Agency. Mr. Grant has extensive experience in the application of systems engineering to the design, development, test and evaluation, and fielding of complex systems involving multisensor integration, command and control, human–machine interfaces, and guidance and control systems. Mr. Grant's engineering leadership in APL prototype systems for the U.S. Navy is now evidenced by capabilities on board more than 100 cruisers, destroyers, and aircraft carriers of the U.S. Navy and its allies. He has served on national committees, including as a technical advisor on studies for the Naval Studies Board of the National Academies of Sciences, Engineering, and Medicine and as a member of the U.S. Strategic Command Senior Advisory Group. He is a member of the program committees for the Department of Electrical and Computer Engineering and the Engineering for Professionals Systems Engineering Program of the Johns Hopkins University Whiting School of Engineering. Mr. Grant earned a BS in physics from the University of Maryland, College Park, and an MS in applied physics and an MS in computer science from the Johns Hopkins University Whiting School of Engineering.

JAMES L. KINTER is the director of the Center for Ocean-Land-Atmosphere Studies (COLA) at George Mason University (GMU), where he oversees basic and applied climate research conducted by the center. Dr. Kinter's research includes studies of atmospheric dynamics and predictability on intra-seasonal and longer time scales, particularly the prediction of Earth's climate using numerical models of the coupled ocean–atmosphere–land system. Dr. Kinter is a tenured professor of climate dynamics in the Department of Atmospheric, Oceanic and Earth Sciences of the College of Science at GMU, where he has responsibilities for teaching climate predictability and climate change. After earning his doctorate in geophysical fluid dynamics at Princeton University in 1984, he served as a National Research Council Associate at the National Aeronautics and Space Administration Goddard Space Flight Center and as a faculty member of the University of Maryland prior to helping to create COLA. A fellow of the American Meteorological Society, Dr. Kinter has served on many national and international review panels for both scientific research programs and supercomputing programs for computational climate modeling. Dr. Kinter has served on three previous National Academies of Sciences, Engineering, and Medicine committees.

RUBY LEUNG is a Battelle Fellow at Pacific Northwest National Laboratory. Her research broadly cuts across multiple areas in modeling and analysis of the climate and water cycle, including orographic precipitation, monsoon climate, extreme events, land surface processes, land–atmosphere interactions, and aerosol–cloud interactions. Dr. Leung is the chief scientist of the Department of Energy's (DOE's) Energy Exascale Earth System Model, a major effort involving more than 100 Earth and computational scientists and applied mathematicians to develop state-of-the-art capabilities for modeling human–Earth system processes on DOE's next-generation, high-performance computers. She has organized several workshops sponsored by DOE, the National Science Foundation, the National Oceanic and Atmospheric Administration, and the National Aeronautics and Space Administration to define gaps and priorities for climate research. Dr. Leung is a member of the National Academies of Sciences, Engineering, and Medicine's Board on Atmospheric Sciences and Climate and an editor of the American Meteorological Society's (AMS's) *Journal of Hydrometeorology*. She has published more than 450 papers in peer-reviewed journals. Dr. Leung is an elected member of the National Academy of Engineering and the Washington State Academy of Sciences. She is also a fellow of AMS, the American Association for the Advancement of Science, and the American Geophysical Union (AGU). She is the recipient of the AGU Global Environmental Change Bert Bolin Award and Lecture in 2019, the AGU Atmospheric Science Jacob Bjerknes Lecture in 2020, and the AMS Hydrologic Sciences Medal in 2022. Dr. Leung was awarded the DOE Distinguished Scientist Fellow in 2021. She received a BS in physics and statistics from the Chinese University of Hong Kong and an MS and a PhD in atmospheric sciences from Texas A&M University.

PARVIZ MOIN is the Franklin P. and Caroline M. Johnson Professor of Mechanical Engineering and the director of the Center for Turbulence Research at Stanford University. He was the founding director of the Institute for Computational and Mathematical Engineering and he directed the Department of Energy's Accelerated Strategic Computing Initiative and Predictive Science Academic Alliance Program centers. Dr. Moin pioneered the use of direct numerical simulation and large eddy simulation techniques for the study of the physics and reduced-order modeling of multiphysics turbulent flows. His current research interests include predictive simulation of aerospace systems, hypersonic flows, multiphase flows, propulsion, numerical analysis for multiscale problems, and flow control. Dr. Moin is the co-editor of the *Annual Review of Fluid Mechanics* and the associate editor of the *Journal of Computational Physics*. Among his awards are the American Physical Society (APS) Fluid Dynamics Prize and American Institute of Aeronautics and Astronautics (AIAA) Fluid Dynamics Award. Dr. Moin is a member of the National Academy of Sciences, the National Academy of Engineering, and the Royal Spanish Academy of Engineering. He is a fellow of APS, AIAA, and the American Academy of Arts and Sciences. Dr. Moin received a PhD in mechanical engineering from Stanford University.

LUCILA OHNO-MACHADO is the deputy dean for biomedical informatics and leads the Section for Biomedical Informatics and Data Science at Yale School of Medicine. Previously, Dr. Ohno-Machado was the health sciences associate dean for informatics and technology, the founding chief of the Division of Biomedical Informatics in the Department of Medicine, and a distinguished professor of medicine at the University of California, San Diego (UCSD). She also was the founding chair of the UCSD Health Department of Biomedical Informatics and founding faculty of the UCSD Halicioğlu Data Science Institute in La Jolla, California. Dr. Ohno-Machado received her medical degree from the University of São Paulo, Brazil; her MBA from the Escola de Administração de São Paulo, Fundação Getúlio Vargas, Brazil; and her PhD in medical information sciences and computer science from Stanford University. She has led informatics centers that were funded by various National Institutes of Health initiatives and by agencies such as the Agency for Healthcare Research and Quality, the Patient-Centered Outcomes Research Institute, and the National Science Foundation. Dr. Ohno-Machado organized the first large-scale initiative to share clinical data across five University of California medical systems and later extended the initiative to various institutions in California and around the country. Prior to joining UCSD, Dr. Ohno-Machado was a distinguished chair in biomedical informatics at Brigham and Women's Hospital and on the faculty at Harvard Medical School and at the Massachusetts Institute of Technology's Health Sciences and Technology Division. She is an elected member of the National Academy of Medicine, the American Society for Clinical Investigation, the American Institute for Medical and Biological Engineering, the American College of Medical Informatics, and the International Academy of Health Sciences Informatics. Dr. Ohno-Machado is a recipient of the American Medical Informatics Association leadership award, as well as the William W. Stead Award for Thought Leadership in Informatics. She serves on several advisory boards for national and international agencies.

COLIN J. PARRIS has achieved significant academic and professional success while attending and leading some of the most prestigious academic and business institutions in the United States and internationally. His career has been centered on the development and enhancement of digital transformation across multiple industries (telecommunications, banking, retail, aviation, and energy) in billion-dollar companies, as well as advocating/evangelizing science, technology, engineering, and mathematics advancement across minority communities. As GE Digital's chief technology officer, Dr. Parris leads teams that work to leverage technologies and capabilities across GE to accelerate business impact and create scale advantage for digital transformation. He also champions strategic innovations and identifies and evaluates new, breakthrough technologies and capabilities to accelerate innovative solutions to solve emerging customer problems. Dr. Parris created and leads the Digital Twin Initiative across GE. He previously held the position of the vice president of software and analytics research at GE Research in Niskayuna, New York. Prior to joining GE, Dr. Parris worked at IBM, where he was an executive for 16 years in roles that spanned research, software development, technology management, and profit and loss management. He was the vice president of system research at the IBM Thomas J. Watson Research Division, vice president of software development for IBM's largest system software development laboratory (more than 6,000 developers worldwide), vice president of corporate technology, and vice president and general manager of IBM Power Systems responsible for the company's more than $5 billion Unix system and software business. Dr. Parris holds a PhD in electrical engineering from the University of California, Berkeley; an MS from Stanford University; an MS in electrical engineering and computer science from the University of California, Berkeley; and a BS in electrical engineering from Howard University.

IRENE QUALTERS serves as the associate laboratory director, emeritus, for simulation and computation at Los Alamos National Laboratory, a Department of Energy national laboratory. Ms. Qualters previously served as a senior science advisor in the Computing and Information Science and Engineering Directorate of the National Science Foundation (NSF), where she had responsibility for developing NSF's vision and portfolio of investments in high-performance computing, and has played a leadership role in interagency, industry, and academic engagements to advance computing. Prior to her NSF career, Ms. Qualters had a distinguished 30-year career in industry, with a number of executive leadership positions in research and development in the technology sector. During her 20 years at Cray Research, she was a pioneer in the development of high-performance parallel processing technologies to accelerate scientific discovery. Subsequently as the vice president, Ms. Qualters led Information Systems for Merck Research Labs, focusing on software, data, and computing capabilities to advance all phases of pharmaceutical research and development.

INES THIELE is the principal investigator of the Molecular Systems Physiology Group at the University of Galway, Ireland. Dr. Thiele's research aims to improve the understanding of how diet influences human health. Therefore, she uses a computational modeling approach, termed constraint-based modeling, which has gained increasing importance in systems biology. Her group builds comprehensive models of human cells and human-associated microbes, then employs them together with experimental data to investigate how nutrition and genetic predisposition can affect one's health. In particular, she is interested in applying her computational modeling approach for better understanding of inherited and neurodegenerative diseases. Dr. Thiele has been pioneering models and methods allowing large-scale computational modeling of the human gut microbiome and its metabolic effect on human metabolism. She earned her PhD in bioinformatics from the University of California, San Diego, in 2009. She was an assistant and associate professor at the University of Iceland (2009–2013) and an associate professor at the University of Luxembourg (2013–2019). In 2013, Dr. Thiele received the ATTRACT fellowship from the Fonds National de la Recherche (Luxembourg). In 2015, she was elected as European Molecular Biology Organization Young Investigator. In 2017, she was awarded the prestigious European

Research Council starting grant. In 2020, she was named a highly cited researcher by Clarivate, and she received the National University of Ireland, Galway, President's award in research excellence. She is an author of more than 100 international scientific papers and a reviewer for multiple journals and funding agencies.

CONRAD TUCKER is an Arthur Hamerschlag Career Development Professor of Mechanical Engineering at Carnegie Mellon University and holds courtesy appointments in machine learning, robotics, biomedical engineering, and the CyLab Security and Privacy Institute. Dr. Tucker's research focuses on employing machine learning/artificial intelligence (AI) techniques to enhance the novelty and efficiency of engineered systems. His research also explores the challenges of bias and exploitability of AI systems and the potential impacts on people and society. Dr. Tucker has served as the principal investigator (PI)/co-PI on federally/non-federally funded grants from the National Science Foundation, the Air Force Office of Scientific Research, the Defense Advanced Research Projects Agency, the Army Research Laboratory, and the Bill and Melinda Gates Foundation, among others. In February 2016, he was invited by National Academy of Engineering (NAE) President Dr. C.D. Mote, Jr., to serve as a member of the Advisory Committee for the NAE Frontiers of Engineering Education Symposium. Dr. Tucker is currently serving as a commissioner on the U.S. Chamber of Commerce Artificial Intelligence Commission on Competitiveness, Inclusion, and Innovation. Dr. Tucker received his PhD, MS (industrial engineering), and MBA from the University of Illinois at Urbana-Champaign and his BS in mechanical engineering from the Rose-Hulman Institute of Technology.

REBECCA WILLETT is a professor of statistics and computer science at the University of Chicago. Her research is focused on machine learning, signal processing, and large-scale data science. Dr. Willett received the National Science Foundation (NSF) CAREER Award in 2007, was a member of the Defense Advanced Research Projects Agency Computer Science Study Group, received an Air Force Office of Scientific Research Young Investigator Program award in 2010, was named a fellow of the Society of Industrial and Applied Mathematics in 2021, and was named a fellow of the IEEE in 2022. She is a co-principal investigator and member of the Executive Committee for the Institute for the Foundations of Data Science, helps direct the Air Force Research Laboratory University Center of Excellence on Machine Learning, and currently leads the University of Chicago's AI+Science Initiative. Dr. Willett serves on advisory committees for NSF's Institute for Mathematical and Statistical Innovation, the AI for Science Committee for the Department of Energy's Advanced Scientific Computing Research program, the Sandia National Laboratories' Computing and Information Sciences Program, and the University of Tokyo Institute for AI and Beyond. She completed her PhD in electrical and computer engineering at Rice University in 2005 and was an assistant professor and then tenured associate professor of electrical and computer engineering at Duke University from 2005 to 2013. She was an associate professor of electrical and computer engineering, a Harvey D. Spangler Faculty Scholar, and a fellow of the Wisconsin Institutes for Discovery at the University of Wisconsin–Madison from 2013 to 2018.

XINYUE YE is a fellow of the American Association of Geographers (AAG) and a fellow of the Royal Geographical Society (with the Institute of British Geographers), holding the Harold L. Adams Endowed Professorship in the Department of Landscape Architecture and Urban Planning and the Department of Geography at Texas A&M University–College Station (TAMU). Dr. Ye directs the focus of transportation in the PhD program of Urban and Regional Science at TAMU and is the interim director of the Center for Housing and Urban Development. His research focuses on geospatial artificial intelligence, geographic information systems, and smart cities. Dr. Ye won the national first-place research award from the University Economic Development Association. He was the recipient of annual research awards in both computational science (New Jersey Institute of Technology) and geography (Kent State University) as well as the AAG Regional Development and Planning Distinguished Scholar Award. He was one of the top 10 young scientists named by the World Geospatial Developers Conference in 2021. His work has been funded by the National Science Foundation, the National Institute of Justice, the Department of Commerce, the Department of Energy, and the Department of Transportation. Dr. Ye is the editor-in-chief of *Computational Urban Science*, an open-access journal published by Springer. He also serves as the co-editor of *Journal of Planning Education and Research*, the flagship journal of the Association of Collegiate Schools of Planning.

G

Acronyms and Abbreviations

AI	artificial intelligence
ASC	Advanced Simulation and Computing
CSE	computational science and engineering
DDDAS	Dynamic Data Driven Application Systems
DoD	Department of Defense
DOE	Department of Energy
DSE	data science and engineering
GUI	graphical user interface
ML	machine learning
MRI	magnetic resonance imaging
NIH	National Institutes of Health
NNSA	National Nuclear Security Administration
NSF	National Science Foundation
OED	optimal experimental design
PDE	partial differential equation
PSAAP	Predictive Science Academic Alliance Program
USGCRP	U.S. Global Change Research Program
VVUQ	verification, validation, and uncertainty quantification